Environmental, Social and Governance Policies, Management Systems & Processes in the Organization, Corporate Governance, and Intellectual Property Rights

Indian Council of
Agricultural Research (ICAR)

Dr. Balasaheb Sawant Konkan
Krishi Vidyapeeth (DBSKKV)

National Agricultural Higher Education
Project-Institutional Development Plan
(NAHEP-IDP)

NIPA® Genx Electronic Resources &
Solutions P. Ltd., New Delhi-110 034

About the Authors

Dr. Kasturi R Naik, Assistant Professor, Sparsh Global Business School| Author| HRBP 2.0 Practitioner | NLP Practitioner | L & D Consultant | Life Coach Specialized in Emotional Intelligence | kasturi.rnaik@sparshgbs.in Link to connect: https://www.linkedin.com/in/dr-kasturi-r-naik-63445614/

Dr. Kasturi R Naik is known for real-time, forthright, solutions to complex people's problems. Knows the difference between Passion, Profession, and Purpose and that has made all the difference to the way she lives her life. She has more than 12 years of academic experience and 5 years of consultancy experience in the fields of human resource management, leadership, and team building for various corporate, educational institutions, and NGOs. At present, she is associated as an Assistant Professor with Sparsh Global Business School. She has been associated as a visiting faculty at JBIMS and GNIMS for MHRD and e-MBA courses and a resource person at XLRI and IIBF. She has also been associated with VMENTOR.ai as a mentor and subject matter expert in the domain of human resource management. Dr. Kasturi R Naik is the recipient of multiple national and international awards and accolades for her enriching work in the field of emotional intelligence, personality enhancement, stress management, and leadership development.

Her credentials include Ph.D. with Stress management as a broad area of research, MMS, BPHARM, DLL, Internationally Certified & Licensed Heal Your Life Workshop Leader and Life Coach, Certified in Positive Psychology: Resilience Skills, Certified in Building Your Leadership Skills by HEC Paris, Certified Executive Life Coach, Certified Image Consultant, Certified Emotional Intelligence Practitioner, Certified NLP Practitioner, and HRBP 2.0 Practitioner. Her research capabilities are self-proven through the 30+ research papers that she has published nationally and internationally, out of which she got the best paper awards in a couple of conferences as well and published in the ABDC category journal.

Dr. Kasturi R Naik has also been honored to play a role of the discussant and session chair in national and international conferences and has also won Best Paper track presenter award at ABRM International conference 2015, besides Best paper award at IMRC 2016.

Areas of Research & Consultancy Interest
Leadership Ecosystem Development
Strategic HR Partnering, and
Learning and Development

Dr. S. Dur Khan, Assistant Professor, JBIMS |MBA (HR), Ph.D| durkhan@jbims.edu Link to connect: https://www.linkedin.com/in/dr-dur-khan-0103bb115/

Dr. Dur Khan is a dedicated educator with extensive experience in the field of higher education. She completed her master's from Welingkar Institute of Management Studies and her Ph.D. from the University of Mumbai. Having started her teaching career with Sydenham Institute of Management Studies, as Faculty in HR, she is currently working as full-time faculty at Jamnalal Bajaj Institute of Management Studies, University of Mumbai.

Her subjects of expertise include Human Resource Management, Organizational Behaviour, Strategic HRM, Performance Management, Learning and Development, and Organization Development. She is the Program Coordinator for MHRD, a two-year full-time Program in Human Resource Management. She is also the professor in charge of CSR and undertakes various social initiatives for the betterment of society.

She is a recognized guide for Ph.D. in Management Studies, University of Mumbai. She has published research papers and cases and presented her work at various national and international research conferences. She is a reviewer for refereed journals and conferences including Elsevier's Journal of Personality and Individual Differences and London School of Management's Online International Conference to name a few.

Beyond teaching, she is a trainer and has facilitated various training programs on talent acquisition, talent management, citizenship behavior, appraisals and 360-degree feedback, training effectiveness, motivation, and empowerment for various public and private sector organizations. She has received a tribute of 'Best Professors in Human Resource Management' by the World HRD Congress, Mumbai.

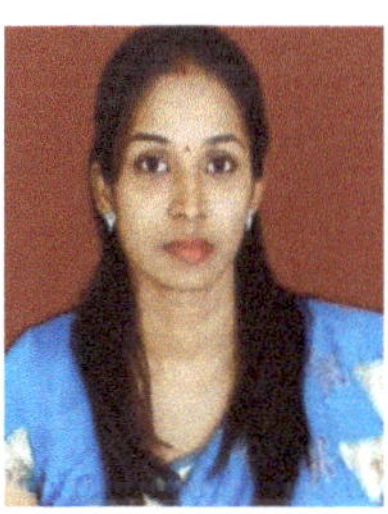

Dr. V. Selvalakshmi, Associate Professor and HOD| Vellammal College of Engineering and Technology|B.Com | D.L.L &L.W | MBA- HR & Marketing | PHD| selvalakshmiv.mba@srmvalliammai.ac.in

Link to connect: https://www.linkedin.com/in/dr-selvalakshmi-bharath-46439018/

A go-getter, Goal oriented, creative and a passionate educationist with 14 years of experience .Have a blend of both corporate and educational experience. At present working as Associate Professor & HOD, Department of Management Studies, Vellammal College of Engineering and Technology. Author have published various research papers both at International and National level.

(SAGE, SCI,Scopus indexed) and also has authored 4 books in management and a e-book which was published by Lap Lambert publishing House, Germany. She has published case studies under Himalaya Publishing House along with renowned academicians from top Bschools. In total she has 3 patents to her credit and have completed 4 ATAL Courses – Design thinking, Strategy, Blended Learning and Flipped Classroom and Leadership & Excellence. She has delivered expert talks as resource person at various institutions. She has presented research papers in various renowned institutions like IIM Indore, IIM Bodhgaya, IIM Trichy, IRMA to name few. She handles various portfolios such as Head – Technology Business Incubator, HeadInstitutional Social responsibility, Core Committee Member for Institution innovation Council, Core committee member – Consultancy Committee, NBA –Committee member and CII (Confederation of Indian Industry) coordinator for College. She was instrumental in building up the Incubation Centre at SRM VEC.She is the Part of Editorial Board member for i-Manager's Journal on Management (JMGT) 2022 to till date. She has completed 6 NPTEL Certification, a MHRD Initiative and certified by IIT Madras and currently pursuing supply chain course from Georgia Institute of Technology, USA.Was Part of Advisory Committee for International Conference on Sustainable Innovation, Research & Emerging Trends in Entrepreneurship, Pratap University, Jaipur. She has coordinated many training programs and has won "Young Researcher" award from Universal innovators leaders in association with National Institute of Technology Patna and University of Valladolid, Spain and Dynamic Teacher (National level) by Avishkar Foundation, Solapur, 2022.

Dr. Anita P Bobade, Professor and I/C Director| DES's Navinchandra Mehta Institute of Technology and Development | B.Com. | MBA | PhD

Email id: anita.bobade@despune.org; anita.bobade@nmitd.edu.in

Link to connect: https://www.linkedin.com/in/dr-anita-bobade-08639811/

Dr. Anita Bobade, is a seasoned finance professional, having worked in Financial Service Sector, for more than 15 years with Institutions like UTI, IDBI Mutual Fund, Asit C Mehta, Sanmar Securities and Merchant Banking outfits. She joined Business School, later in life. Presently she is working as a Professor, MMS, with DES's Navinchandra Mehta Institute of Technology and Development and is in academics for 2 decades. Dr. Anita is an approved Professor and PhD guide of University of Mumbai. Dr. Anita Bobade, has been recently promoted as an I/C Director of DES NMITD. Dr. Anita is an established author, blogger and Financial Literacy Activist.

Recently she has authoured a book on Amazon, titled, " How You can Make a Million on the Stock Market" She has also translated this in Marathi, शेर मार्केट–तंत्र आणि मंत्र

She is a B. Com of University of Mumbai, MBA from CSIBER and PhD from Leeds University, United Kingdom, in the area of Strategic Management. Dr. Anita is presently pursuing an interdisciplinary research, as a Post Doc Student of Oxford Center for Hindu Studies (OCHS) United Kingdom. The proposed title of the thesis is "Vedanta, Dharma and Management." Dr. Anita is also an approved PhD Guide for University of Mumbai in Management Studies. Dr. Anita has authored over 32 research papers in peer reviewed journals and have 16 books, 32 research papers to her credit for under graduate and post graduate students of Management at University of Mumbai and two chapters in International books published by Palgrave Macmillan, U.K.

She is also a reviewer for Scopus Based Inderscience Journal and International Journal of Business and Economics Research. Her areas of research interest: Strategic Management, Perspective Management, Financial Management and Financial Markets and Institution. Dr. Anita is nominated as a Country Director, for EuroMed Research & Business Institute, Switzerland since 2016 to promote their publications and conferences in India.

Her other notable achievements include, Member of the University of Mumbai, Sesquicentennial Committee, which published a souvenir which is a rare and unique collection of Art and Literature, applauded by the then President of India, His Excellency, Hon. Dr. APJ Kalam.

Dr. Anita is the recipient of Global Digital Teacher Award in November 2020, given by Global Digital Academy for the Ongoing Dedication of Impacting Excellence Through the Development and Dissemination of Radically Transforming Learning Spectrum during Covid Pandemic 2019, through her online certifications and webinars for students and housewives to spread financial literacy called, " Dhan Ki Baat".

Dr. Anita Bobade's Social media handles

https://www.linkedin.com/in/dr-anita-bobade-08639811/
https://www.researchgate.net/profile/Anita-Bobade/savedlist
http://dranitabobade.com/
https://www.amazon.in/dp/B092ZDMKCB
https://www.amazon.in/dp/B08LPN4WQD
https://indianethosinmanagement.wordpress.com/
https://www.youtube.com/channel/UC-9K64MsSPxnvcIfTHbQPKg
https://www.facebook.com/anita.bobade.144/

Environmental, Social and Governance Policies, Management Systems & Processes in the Organization, Corporate Governance, and Intellectual Property Rights

Compiled by

Prashant S. Bodake, Dean & Director of Instruction, DBSKKV
Co-ordinator & Principal Investigator, NAHEP-IDP, DBSKKV-Dapoli

Srinivasan Iyengar, Director of Jamnalal Bajaj Institute of Management Studies, Mumbai

Vinayak K. Patil, Co- Principal Investigator, NAHEP-IDP, DBSKKV-Dapoli

Authors

Kasturi Naik, Head &Assistant Professor,Sparsh Global Business School Greater Noida, Delhi

Dur Khan, Associate Professor, J.B.I.M.S., University of Mumbai

V. Selvalakshmi, Assistant Professor, Head of Incubation Centre SRM Valliammai Engineering College, Kanchipuram

Anita Bobade, Professor, DES Navinchandra Mehta Institute of Technology & Development, Dadar-Mumbai

Indian Council of Agricultural Research (ICAR)

Dr. Balasaheb Sawant Konkan Krishi Vidyapeeth (DBSKKV)

National Agricultural Higher Education Project-Institutional Development Plan (NAHEP-IDP)

NIPA® Genx Electronic Resources & Solutions P. Ltd., New Delhi-110 034

Citation: National Agriculture Higher Education Project Institutional Development Plan, DBSKKV, Dapoli & ICAR, New Delhi

NIPA® GENX ELECTRONIC RESOURCES & SOLUTIONS P. LTD.
101,103, Vikas Surya Plaza, CU Block
L.S.C.Market, Pitam Pura, New Delhi-110 034
Ph : +91 11 27341616, 27341717, 27341718
E-mail: newindiapublishingagency@gmail.com
Website: www.nipabooks.com

For customer assistance, please contact
Phone: + 91-11-27 34 17 17
Fax: + 91-11-27 34 16 16

ISBN: 978-93-58871-03-6

Composed and Designed by NIPA®.

भारतीय कृषि अनुसंधान परिषद
कृषि अनुसंधान भवन–II, नई दिल्ली–110012
INDIAN COUNCIL OF AGRICULTURAL RESEARCH
KRISHI ANUSANDHAN BHAVAN-II, PUSA, NEW DELHI-110012
Phone: +91-11-25841760
Email: ddg.edu@icar.gov.in, ddgedn@gmail.com, nd.nahep@icar.gov.in

Dr. Rakesh Chandra Agrawal
Deputy Director General (Agril. Edn.)

Message

It is with immense pleasure and a sense of purpose that I provide this forward to the National Agricultural Higher Education Project-Institutional Development Plan (NAHEP-IDP) at Dr. Balasaheb Sawant Konkan Krishi Vidyapeeth (DBSKKV) and the concurrent endeavour to develop a short pilot course in skilling small-scale agri-entrepreneurship. This transformative initiative is under the guidance and leadership of the Hon'ble Vice Chancellor of DBSKKV Dr.S.G.Bhave.

Principal Investigator of NAHEP-IDP, DBSKKV-Dapoli Dr. P.S.Bodake, a distinguished and visionary leader in the field of agriculture, has consistently championed the cause of advancing agricultural education, research, and extension services. His strategic vision has played a pivotal role in shaping the trajectory of agricultural development in our nation, with a keen focus on empowering farming communities and fostering innovation.

The NAHEP-IDP of DBSKKV stands as a testament to ICAR & World Bank commitment to elevating the standards of agricultural education. By providing institutional support and promoting excellence across various facets of academia, research, and extension activities, the project aims to position DBSKKV as a center of excellence in agricultural learning and research.

In tandem with the NAHEP-IDP, the initiation of a short pilot course in skilling smallscale agri-entrepreneurship reflects DBSKKV foresight in addressing contemporary challenges in agriculture. Recognizing the pivotal role played by small-scale agri-entrepreneurs in driving rural economies, this course is designed to equip them with the necessary skills and knowledge to navigate the dynamic landscape of agribusiness.

The NAHEP-IDP, DBSKKV envisions this pilot course as a stepping stone toward fostering a culture of entrepreneurship in agriculture. By blending theoretical insights with practical applications, the course seeks to empower agri-entrepreneurs, enabling them to adopt sustainable farming practices, embrace innovation, and contribute significantly to the economic prosperity of their communities.

I extend my heartfelt appreciation to Dr. P.S. Bodake and his entire NAHEP-IDP Team for their steadfast leadership and unwavering commitment to advancing agricultural education and entrepreneurship. Under his guidance, these initiatives are poised to make a lasting impact on the agricultural education landscape and the entrepreneurial spirit within the students and farming community.

(R.C. Agrawal)
Deputy Director General (Agril. Edn.)

DR. BALASAHEB SAWANT KONKAN KRISHI VIDYAPEETH
Dapoli, Dist. Ratnagiri - 415712
Tel. No.: (0)02358-282064, **Fax No.**: 02358-282074, **Mob.**: 9422556565
E-mail : vcdbskkv-mh@gov.in, ipmbhave@gmail.com, **website** : www.dbskkv.org

Dr. Sanjay G. Bhave
IDP Leader
Vice Chancellor, DBSKKV

Message

I am delighted to present the Institutional Development Plan as part of the National Agricultural Higher Education Project (NAHEP) undertaken by our esteemed institution. This initiative marks a significant milestone in our commitment to advancing agricultural education, research, and innovation, with the ultimate goal of transforming the agricultural landscape of our nation.

This pilot course serves as a testament to our university's dedication to fostering innovation and fostering collaboration between academia and industry. We believe that by offering this course, we are not only contributing to the skill development of individuals but also playing a pivotal role in shaping the future of agriculture.

We, at DBSKKV are happy to felicitate the proficient attainment of the Principal Investigator, IDP Coordinator, and Dean and Director of Instruction of DBSKKV- Dr.Prashant S. Bodake and team NAHEP-IDP involved in the same vertical. Your magnificent work makes all of us highly gratified and prosperous. Well Done!

I extend my gratitude to the faculty members, industry experts, and stakeholders who have contributed their expertise and insights to the development of this course. Their collective efforts have ensured that the curriculum aligns with the ever-changing demands of the agricultural sector.

I am excited to witness the impact that this short pilot course on skilling small-scale agricultural entrepreneurship will have on our participants and, consequently, on the agricultural landscape. With this initiative, we are sowing the seeds of empowerment and innovation, cultivating a new generation of agripreneurs who will not only navigate challenges but also unearth opportunities for growth.

(Sanjay G.Bhave)
IDP Leader
Vice Chancellor, DBSKKV

NATIONAL AGRICULTURAL HIGHER EDUCATION PROJECT
AGRICULTURAL EDUCATION DIVISION
INDIAN COUNCIL OF AGRICULTURAL RESEARCH
(Department of Agricultural Research and Education
Ministry of Agriculture and Farmers Welfare, Government of India)
Room No. 514, Krishi Anusandhan Bhawan-II, Pusa, New Delhi-110 012

Date : 14.12.2023

Dr. N.K. Jain
National Coordinator (IDP)

Message

It is a matter of great pleasure to learn that Dr. Balasaheb Sawant Konkan Krishi Vidyapeeth, Dapoli has developed short pilot course on "Skilling Small-scale Agri-Entrepreneurship" for the undergraduate students. The efforts made by DBSKKV, Dapoli and the Knowledge Experts is a significant stride towards fostering interdisciplinary learning and preparing students for the evolving landscape of agriculture. Agriculture, the backbone of our nation, is undergoing a transformative phase that demands a comprehensive understanding of not only traditional farming practices but also entrepreneurial skills. In this context, DBSKKV, Dapoli is driven by the vision to equip undergraduate students with the knowledge and skills necessary to navigate the complexities of modern agriculture and contribute to the burgeoning field of agri-entrepreneurship. I commend the leadership of Dr. S.G. Bhave, Hon'ble Vice Chancellor, DBSKVV, Dapoli for his forward-thinking approach in embracing this initiative and commitment to provide a robust academic environment aligns seamlessly with our objective of preparing students for the dynamic and competitive landscape of agriculture and entrepreneurship.

I also congratulate Dr. P.S. Bodake, Principal Investigator of NAHEP-IDP, DBSKKV, Dapoli and his entire team, whose collaborative efforts have been instrumental in shaping the curriculum of this course. Together, we envision a program that not only imparts knowledge but also nurtures a spirit of inquiry and enterprise among the students. I extend my warmest congratulations to the University on the launch of the innovative short pilot course on "Skilling Small-scale Agri-Entrepreneurship". May this endeavour be a beacon of progress, fostering excellence in agricultural education and empowering the next generation of agri-entrepreneurs. University's commitment to bridging theory and practical is commendable, and I am confident that this initiative will contribute significantly to the growth of our agricultural landscape.

Place: New Delhi

(N. K. Jain)

Mob. No.: 9420413255 / 8275201130
Email.: (P)prameghash@gmail.com (O) hodagrodapoli@gmail.com (O) piidpdbskkv@gmail.com (O)nodalcelldbskkv@gmail.com

Dr. P. S. Bodake
Dean & Director of Instruction
Coordinator, Principal Investigator, NAHEP-IDP

Message

It is with immense pride and anticipation that I present the Institutional Development Plan as part of our active involvement in the National Agricultural Higher Education Project (NAHEP). This plan embodies our institution's unwavering commitment to redefining agricultural education and research, and it signifies a transformative journey towards excellence and innovation.

Through this plan, we aim to enhance our institutional capacities, bolster research initiatives, and cultivate strategic partnerships with industries and communities. By incorporating modern pedagogical techniques, fostering interdisciplinary collaboration, and harnessing the potential of emerging technologies, we strive to create a learning ecosystem that empowers our students to become leaders and change-makers in the agricultural sector and it will not only elevate our institution's stature but also propel the agricultural education and research landscape towards a brighter and more sustainable future.

Our joint commitment to education and empowerment has culminated in the creation of this pilot course. Through a harmonious blend of theoretical insights and practical experiences, participants will delve into market trends, value chain analysis, sustainable practices, financial management, and the integration of technology into agriculture. This holistic approach is designed to prepare individuals to navigate the intricacies of agripreneurship with confidence.

This pilot course is an embodiment of our dedication to fostering collaboration between academia and industry. By bridging the gap between theory and practice, we aim to nurture a new generation of agripreneurs who are poised to steer the agricultural sector towards innovation and growth.

We sincerely mention primitive gratitude towards the former Vice-chancellor of DBSKKV Dr. Sanjay D. Sawant and Vice-chancellor of DBSKKV Dr. Sanjay G. Bhave for such prudent leadership and providence. We feel proud to highlight the sheer hard work and great dedication of the Co-Principal Investigator of NAHEP-IDP and IDP Staff.

We are grateful to the Director of Jamnalal Bajaj Institute of Management-Mumbai Dr. Shrinivasan Iyengar for such excellent association and cordial congruence in the course of developing a short pilot course in skilling small-scale Agri-Entrepreneurship. We also express our sincere gratitude to the faculty members of JBIMS & University of Mumbai, industry experts, and stakeholders whose expertise and contributions have shaped this course. Their collective wisdom ensures that the curriculum remains attuned to the evolving demands of the agricultural landscape.

Thank you.

(P. S. Bodake)
Dean & Director of Instruction
Coordinator, Principal Investigator, NAHEP-IDP

Preface

We are delighted to present to you our comprehensive book titled "Environmental, Social and Governance Policies: Management Systems & Processes in the Organization, Corporate Governance, and Intellectual Property Rights." This book is a valuable resource for individuals seeking to delve into the intricate world of sustainability, governance, and intellectual property rights within the context of Indian agriculture.

Authored by a team of educators, this book brings together the expertise and insights of Dr. Kasturi R. Naik, an Assistant Professor at Sparsh Global Business School; Dr. Dur Khan, an Assistant Professor at JBIMS; Dr. V Selvalakshmi, an Associate Professor and HOD at Vellammal College of Engineering and Technology; and Dr. Anita Bobade, the I/C Director and Professor at DES's NMITD. We would like to express our sincere gratitude to Dr. Srinivasan R. Iyengar, Director of JBIMs, and Dr. P.S. Bodake, NAHEP-IDP Coordinator, for their invaluable contributions to this book project. Dr. Iyengar's visionary leadership and insightful guidance, combined with Dr. Bodake's meticulous coordination, have been instrumental in shaping the content and success of this endeavor.

Their collective knowledge and extensive experience in the fields of environmental policies, social policies, corporate governance, and intellectual property rights bring a unique perspective to this book.

The contents of this book dive into crucial areas:

1. Environmental Policies and Management Systems & Processes in Indian Agriculture Context: This section examines the environmental challenges faced by the agricultural sector and explores effective policies and management systems that can foster sustainable practices.
2. Social Policies and Management Systems & Processes in Indian Agriculture Context: Here, the author address the social aspect of agriculture, focusing on policies and management systems that promote social responsibility, farmer welfare, and community engagement.
3. Intellectual Property Rights in Indian Agriculture Context: This section delves into the complexities of intellectual property rights in the agricultural

landscape, discussing the importance of protecting innovations, patents, and trademarks to drive growth and development.

4. Corporate Governance and Sustainability Reporting in Indian Agriculture Context: This additional section delves into the realm of corporate governance and sustainability reporting, shedding light on the significance of transparent and ethical practices for the agriculture sector.

With a blend of theoretical frameworks, practical case studies, and real-world examples, this book offers a comprehensive guide for students, educators, professionals, researchers, and policy-makers, enabling them to understand and navigate the interplay between environmental sustainability, social responsibility, corporate governance, and intellectual property rights within the Indian agricultural context.

Embark on this enlightening journey as we uncover the critical aspects of sustainable agriculture and the policies that pave the way for a greener and more socially conscious future.

Contents

Message by DDG (Agril. Edn.) Dr. Rakesh Chandra Agrawal .... *ix*

Message by Vice Chancellor Dr. Sanjay G. Bhave .... *xi*

Message by National Coordinator (IDP) Dr. N.K. Jain .... *xiii*

Message by Principal Investigator (NAHEP-IDP) Dr. Prashant S. Bodake .... *xv*

Preface .... *xvii*

Environmental, Social and Governance Policies, Management Systems & Processes in the Organization, Corporate Governance, and Intellectual Property Rights .... *xxiii*

1. Environmental Policies and Management Systems & Processes in Indian Agriculture Context **1**

Learning Objectives 1

1.1 Overview of the Current state of Environmental Policies in Indian Agriculture and Challenges Faced 2

1.2 The Importance of Implementing Environmental Policies and Management Systems in Indian Agriculture 4

1.3 Overview of Key International Standards Such as ISO 14001 and EMAS, and their Requirements in the Context of Indian Agriculture 7

1.4 Best Practices and Case Studies of Companies and Organizations that have Successfully Implemented Environmental Policies and Management Systems in Indian Agriculture 9

1.5 The Challenges Faced in Implementing and Adhering to Environmental Policies and Management Systems in Indian 12

References *19*

2. Social Policies and Management Systems & Processes in Indian Agriculture Context **23**

Learning Objectives 23

1.1 Overview of the Indian Agriculture Sector and Challenges Faced ... 23

1.2 Overview of the Current State of Social Policies in Indian Agriculture ... 27

1.3 Overview of Management Systems & Processes in the Indian Agriculture Context... 31

1.4 Overview of Key International Standards such as ISO 26000 and SA8000, and their requirements in the context of Indian Agriculture ... 33

1.5 Best Practices and Case Studies of Successful Implementation of Social Policies and Management Systems in Indian Agriculture . 34

1.6 The Way Forward for Policy Makers... 37

References ... *43*

3. Intellectual Property Rights in Indian Agriculture Context ... 45

Learning Objectives ... 45

1.1 Introduction to Intellectual Property ... 45

1.2 Current Statutory IP Laws in India vis-vis Agri-based Technologiesin India ... 49

1.3 Benefits of IPR ... 53

1.4 Challenges... 54

1.5 Classification of IPR and its Implications in Agriculture ... 55

1.6 Overview of Indian Agriculture...58

1.7 Protection of Plant Varieties and Farmers' Rights Act (PPV & FR), 2001 ... 59

1.8 History of the Protection of Plant Varieties and Farmers' Rights Act... 59

1.9 Objectives of the Protection of Plant Varieties and Farmers' Rights Act, 2001 ... 60

1.10 Registration of Varieties... 61

1.11 International Union for Protection of New Plant Varieties (UPOV Convention) ...62

References ... *67*

4. Corporate Governance and Sustainability Reporting in Indian Agriculture Context ... 69

Learning Objectives ... 69

1.1 Introduction... 69

1.2 Need For Corporate Governance ... 70

1.2.a Objectives of Corporate Governance.. 70

1.2.b Definitions of Corporate Governance 70

1.2.c Principles of Corporate Governance... 71

1.3 History and Evolution of Corporate Governance in UK 72

1.4 Indian Model of Corporate Governance... 74

1.4.a Corporate Governance – Anglo Saxon (American) Model....... 77

1.4.b Corporate Governance – German and Japanese Model............ 77

1.4.c Corporate Governance – India – Private Sector 77

1.4.d Corporate Governance – India – Public Sector 78

1.4.e Corporate Governance – India – Co-Operative Sector............. 78

1.5 Parties to Corporate Governance in A Co-Operative Society/Bank ... 79

1.6 Corporate Governance in Indian Co-operative Banks 79

1.7 The Global Reporting Initiative (GRI) 81

1.7.a. Background on Sustainability Reporting 81

1.7.b NAHEP 82

1.7.c ICAR 82

1.7.d. GRI, or Global Reporting Initiative 82

Questions.. 85

References *86*

Annexures 87

Annexure 1: Recap of Key Points Covered in the Volume 87

Annexure 2: How Implementing Environmental, Social, and Governance Policies, Management Systems, and Intellectual Property Rights Can Lead to Long-Term Sustainable Growth and Development in Indian Agriculture? 87

Annexure 3: Call-to-Action for Organizations to Take Action on These Important Topics in the Context of NAHEP and Indian Agriculture 87

Environmental, Social and Governance Policies, Management Systems & Processes in the Organization, Corporate Governance, and Intellectual Property Rights

Introduction to the Volume

Overview of the Indian Agricultural Sector and the Need for Sustainable Growth and Development

India's agriculture sector is the backbone of its economy, contributing significantly to the gross domestic product (GDP) and employing over 42.7% of the country's population(World Bank, 2020). Despite the sector's important role, it faces several sustainability challenges that threaten its long-term growth potential. Climate change, soil degradation, water scarcity, and declining soil fertility are a few of the critical risks that the sector is grappling with. Therefore, ensuring sustainable growth and development in the Indian agricultural sector is crucial.

Explanation of the Role of Environmental, Social, and Governance Policies, Management Systems & Processes in the Organization, Corporate Governance, and Intellectual Property Rights in Achieving Sustainability in Agriculture

Environmental, social, and governance (ESG) policies, management systems & processes, and corporate governance play a vital role in achieving sustainability in agriculture. Environmental policies ensure that agricultural practices are environmentally sustainable and natural resources are used effectively. Social policies ensure that the farmer's well-being and rural communities are taken care of while governance policies help in effective decision-making to ensure that agriculture is sustainable in the long term. Management systems & processes help in the effective implementation of these policies. Intellectual property rights promote the innovation and development of new agricultural practices and technologies that are environmentally friendly and sustainable.

Introduction to the National Agricultural Higher Education Project (NAHEP) and Its Goals for Sustainable Agriculture

The National Agricultural Higher Education Project (NAHEP) is a flagship initiative of the Government of India aimed at enhancing the quality of higher education in agriculture and promoting sustainable agriculture. The project has several goals, including providing financial support to higher educational institutions and research institutions to strengthen infrastructure, improve curricula, and promote research. NAHEP seeks to improve the agricultural education system in India and create a better educational infrastructure for agriculture.

Purpose of the Volume and What Readers Can Expect to Learn from It

The primary purpose of this volume is to provide readers with an understanding of the role of environmental, social, and governance policies, management systems & processes, corporate governance, and intellectual property rights in achieving sustainable growth and development in the Indian agriculture sector. Readers can expect to learn about the various sustainability challenges faced by the Indian agricultural sector and how ESG policies and management systems & processes are essential for overcoming these challenges. Additionally, they will learn the importance of corporate governance and intellectual property rights in creating a conducive environment for sustainable agriculture.

Furthermore, this volume will introduce readers to the NAHEP and other initiatives and their goals for promoting sustainable agriculture. As the project has a specific emphasis on enhancing higher education, the volume will detail how improving the quality of agricultural education can positively impact the sector's sustainability. Readers can expect to gain insights into how the academic sector can contribute to the achievement of sustainable agriculture goals in India by collaborating with stakeholders from the government, private sector, and civil society. Overall, the volume aims to create awareness of the need for sustainable agriculture and highlight the role of ESG policies, management systems & processes, corporate governance, and intellectual property rights in achieving this goal.

1

Environmental Policies and Management Systems & Processes in Indian Agriculture Context

Kasturi R. Naik

Assistant Professor, Sparsh Global Business School

Learning objectives

1-1. Identify the current state of environmental policies in Indian agriculture and the challenges faced in their effective implementation.

1-2. Evaluate the importance of implementing environmental policies and management systems in Indian agriculture to ensure sustainable development.

1-3. Analyze key international standards, such as ISO 14001 and EMAS, andhow they can be applied in the context of Indian agriculture.

1-4. Compare and contrast best practices and case studies of companies and organizations that have successfully implemented environmental policies andmanagement systems in Indian agriculture to draw insights and lessons learntfrom them.

1-5. Design and propose solutions to overcome the challenges faced in implementing and adhering to environmental policies and management systems in Indian agriculture.

Agriculture is an essential sector in India, but it has a significant impact on the environment. The Indian government is taking various steps to mitigate the environmental challenges associated with agriculture and promote sustainable practices. To achieve these goals, the National Mission on Sustainable Agriculture (NMSA) aims to reduce greenhouse gas emissions, enhance soil health, and conserve natural resources.

Another significant program is the Rainfed Area Development Program (RADP), which promotes soil and water conservation in rainfed areas. The government is also introducing various certification systems to promote organic farming and reduce the use of chemical fertilizers. One such initiative is the National Agricultural Higher Education Project (NAHEP), which aims to promote innovation, entrepreneurship, and skill development in agriculture. It has a sub-component, i.e., the Institutional Development Plan (IDP) for improving the infrastructure and facilities of agricultural universities and research institutions. The government is also promoting the use of technology in agriculture through the National Agricultural Technology Project (NATP). It fosters research and development of drought-resistant crops, which can reduce the environmental risks associated with farming in water-scarce areas.

Overall, India is actively promoting sustainable agricultural practices through various policies and management systems, while also promoting innovation and skill development in the agriculture sector.

1.1. Overview of the Current state of Environmental Policies in Indian Agriculture and Challenges Faced

Agriculture is a vital sector in the Indian economy, contributing to over 17% of the country's Gross Domestic Product (GDP) and employing more than 50% of its workforce. However, it also accounts for a significant proportion of environmental degradation, including deforestation, soil erosion, and water contamination. In recent years, the Indian Government has implemented several policies aimed at promoting sustainable agriculture practices and minimizing environmental impacts, but significant challenges remain.

One of the essential policies of the Government of India is the National Action Plan for Climate Change (NAPCC), launched in 2008. The NAPCC aims to develop a comprehensive and integrated approach to address climate change challenges by focusing on eight missions, one of which is the National Mission on Sustainable Agriculture (NMSA). The NMSA aims to enhance the resilience of Indian agriculture to climate change impacts by promoting sustainable land use practices, water conservation, and enhancing soil health. Under the NMSA, several schemes like the Paramparagat Krishi Vikas Yojana, Soil Health Card Scheme, and Pradhan Mantri Fasal Bima Yojana are implemented to promote sustainable farming practices.

Another critical policy initiative is the Soil Health Card (SHC) Scheme, launched in 2015. The SHC Scheme aims to provide farmers with a comprehensive analysis of their soil health and recommend appropriate measures to maintain and improve fertility. So far, the scheme has covered more than 23 crore soil samples, and the government plans to cover all farmers by 2022. The scheme is an essential tool for farmers to adopt sustainable farming practices, use fertilizers judiciously, and reduce pollution.

Despite these policy initiatives, numerous challenges are hindering the adoption of sustainable agriculture practices and environmental protection in India. One such challenge is the small landholding size of farmers, leading to limited adoption of new farming practices and technologies. Another challenge is the over-dependence on chemical fertilizers, leading to soil degradation and water pollution. Besides, erratic weather patterns and the loss of agro-biodiversity are further exacerbating the challenges for Indian agriculture.

In recent years, the Indian Government has implemented several policies to encourage sustainable agriculture practices and minimize the environmental impact of farming. These policies have increased awareness and motivated farmers to adopt sustainable practices, but numerous challenges remain. One of the latest policies aimed at incentivizing sustainable agriculture practices is the Farmers' Produce Trade and Commerce (Promotion and Facilitation) Act, 2020. This act creates a legal framework to facilitate trade and commerce of farmers' produce outside of the physical boundaries of APMC markets where farmers can receive better prices for their products. It promotes direct marketing and reduces farmers' reliance on intermediaries to access markets, improving profits and promoting sustainable agriculture practices. Another policy initiative is the implementation of the Pradhan Mantri Fasal Bima Yojana (PMFBY) in 2016. PMFBY provides crop insurance to farmers, aiming to reduce the financial risk associated with farming and promote sustainable practices. The scheme ensures farmers receive compensation for crop loss due to natural calamities. It makes farmers less vulnerable to the risks of farming and encourages them to invest a part of their earnings in environmental conservation. However, the scheme has faced implementation challenges, and the benefits are not reaching the targeted farmers.

The government has also been promoting precision farming practices to reduce input costs and make crop cultivation efficient. Precision farming takes advantage of computerized mapping and data analytics to apply precise amounts of fertilizer and water to crops as and when required, resulting in yield optimization. The adoption of precision farming increases agricultural production while minimizing resource waste and negative environmental impacts of over-fertilization. However, precision farming has a significant initial investment and requires technical skills, making it challenging to adopt by small and marginalized farmers.

Despite various policies, Indian agriculture still faces several environmental challenges. One of the most significant challenges is soil degradation due to the overuse of chemical fertilizers and pesticides. Intensive farming practices have depleted the soil's nutrients and altered its biological characteristics, leading to erosion, nutrient loss, and water contamination. The government has taken several steps, like promoting organic farming and soil health cards, to mitigate the impact of soil degradation. However, the adoption of sustainable farming practices remains a challenge, particularly for small and marginalized farmers, due to the lack of knowledge and resources. Another challenge facing Indian agriculture is the over-dependence on irrigation water from groundwater reservoirs. The overuse of groundwater has led to depletion of the water table, making it increasingly difficult for farmers to access the required amount of water to irrigate their crops. This has led to low crop yields and increased competition for water resources, leading to frequent conflicts. The government has taken initiatives such as the Pradhan Mantri Krishi Sinchai Yojana to promote water conservation and efficient irrigation practices. Still, effective water management requires more sustainable water management practices, such as rainwater harvesting, micro-irrigation, and community-based water management.

The Indian Government has taken several policy initiatives in recent years to promote sustainable agriculture practices and save the environment from the negative effects of farming. However, several challenges, such as soil degradation, water management, and adoption of new technologies, require persistent attention. The government needs to focus on increasing the adoption of sustainable agricultural practices, promoting education and awareness about environmental conservation and partnering with farmers and the private sector to scale-up innovations that improve agricultural productivity and protect the environment.

1.2. The Importance of Implementing Environmental Policies and Management Systems in Indian Agriculture

Environmental policies and management systems are crucial for Indian agriculture to ensure sustainable food production, environmental conservation, and long-term economic growth. Agriculture is the backbone of the Indian economy employs over 50% of the workforce and contributes to around 17% of the country's GDP. It also plays an essential role in global food security, as India is one of the world's largest food producers. India faces several environmental challenges due to intensive farming practices, including soil degradation, water scarcity, and chemical contamination. Environmental policies and management systems are necessary to address these challenges while promoting sustainable food production and rural development.

Effective environmental policies and management systems are essential for the environmental conservation of Indian agriculture. Unsustainable farming practices, such as excess use of chemical fertilizers, pesticides, and groundwater irrigation, have led to several environmental problems. Over- dependence on water resources for irrigation leads to a depletion of water resources and affects local ecosystems and biodiversity. Poor soil management leads to soil degradation, reducing soil fertility and productivity, and restricting the capacity of the soil to sequester carbon. Overuse of chemical fertilizers and pesticides also harm ecology by entering into the food chain and causing pollution in water sources. It is crucial to have effective policies making environmentally friendly farming practices mandatory to prevent further environmental degradation

Implementing environmental policies and management systems can also improve the long-term economic prospects of Indian agriculture. Sustainable agriculture practices, such as crop diversification, organic farming, and rainwater harvesting, conserve environmental resources, reduce the cost of inputs, and promote long-term productivity. Farmers can save up to 30% of their production costs by adopting conservation measures, increasing their profits and overall economic growth. In addition, these practices help avoid environmental disasters such as soil erosion, floods, and droughts that cost billions of dollars in economic losses. Emphasis on sustainable practices guarantees higher food production and preservation of ecological resources for future generations to come

Moreover, by adopting environmental policies and management systems, India can fulfill its international commitments toward environmental conservation and combat climate change. According to the Paris Agreement, India committed to reducing its emission intensity by 33-35% from 2005 levels by 2030. Agriculture is a significant contributor to carbon emissions from developing countries, accounting for up to 23% of total emissions in South Asia. Reducing emissions through sustainable farming practices, such as promoting green manure and reducing tillage, would help India achieve its environmental goals and help combat global climate change. Furthermore, India is committed to achieving the Sustainable Development Goals (SDGs), which include targets for environmental conservation and sustainable food production. Implementing environmental policies and management systems would help achieve these goals

Implementing environmental policies and management systems in Indian agriculture is critical for the country's environmental and economic development. Sustainable farming practices promote resource conservation, long-term economic growth, and fulfilment of international commitments to environmental conservation and cooperation. An agricultural sector that prioritizes environmental concerns is crucial for India's wealth, food security, and environmental health, and can lead the way towards prioritizing sustainability over short-term gains (Singh et.al, 2017; Das et.al, 2019; Mahapatra et.al, 2019; Kumar et.al, 2020)

One of the key challenges of Indian agriculture is the unsustainable use of natural resources. The indiscriminate use of chemical fertilizers and pesticides has depleted soil fertility and caused soil erosion. The overexploitation of groundwater for irrigation has also led to depletion of water resources. Furthermore, intensive farming practices have led to biodiversity loss, which affects ecosystems' functioning. Therefore, it is crucial to implement policies and management strategies that promote sustainable use of natural resources. The Indian government has initiated several policies and programs to address the environmental challenges of agriculture. The National Mission on Sustainable Agriculture (NMSA) aims to promote sustainable agriculture practices and conserve natural resources. It encourages the use of organic farming practices, agroforestry, and efficient water use to increase productivity sustainably. The NMSA also focuses on minimizing greenhouse gas emissions, which is essential to mitigate the effects of climate change on agriculture. Another critical program is the Rainfed Area Development Program (RADP), which aims to improve soil and water conservation in dryland areas. It encompasses practices such as watershed development, groundwater recharge, and soil conservation. The program has witnessed significant success, with farmers in drylands reporting increased crop yields and better incomes. The government has also introduced various certification systems that certify the products are grown using sustainable practices, thereby reducing the environmental impact of agriculture. For example, the National Program for Organic Production (NPOP) promotes organic farming practices, which avoid the use of synthetic fertilizers and pesticides. Organic farming practices help to improve soil health, minimize water use, and promote biodiversity, which is essential for sustainable development.

The Indian government is also promoting the use of information and communication technology (ICT) in agriculture. The e-National Agricultural Market (e-NAM) facilitates the transparent online trading of agricultural products, which helps farmers access information about fair market prices. The use of online tools like soil health cards and mobile apps for farmers to access information regarding weather, soil health, and pests/diseases control can also promote sustainable agriculture practices. One of the critical components of sustainable agricultural development is the promotion of research and development (R&D). The National Agricultural Technology Project (NATP) fosters the development of new technologies and research for drought-resistant crops, which can help overcome environmental risks. Implementing sustainable agricultural practices requires the involvement of all stakeholders, including governments, farmers, and the private sector. Additionally, the government has launched the National Agricultural Higher Education Project (NAHEP) for promoting innovation, entrepreneurship, and skill development in agriculture, particularly at the institutional level. The program aims to provide the necessary education, training, and scientific expertise to promote the use of the latest technologies in farming.The implementation of environmental policies and management systems is essential for sustainable development in Indian agriculture. The sustainable management of natural resources, the reduction of greenhouse gas emissions, and the promotion of organic farming practices can improve agricultural productivity and promote sustainable agriculture. The government's initiatives like NMSA, RADP, and NATP, combined with the promotion of eco-friendly farming practices among farmers, can ensure that agriculture contributes to sustainable development in India.

1.3. Overview of Key International Standards Such as ISO 14001 and EMAS, and their Requirements in the Context of Indian Agriculture

ISO 14001 and EMAS are key international environmental management standards that can be applied to Indian agriculture to promote sustainable farming practices that are in compliance with environmental regulations. ISO 14001 is a globally recognized environmental management standard that provides a systematic approach to developing and implementing environmental management systems (EMS) that ensure the environmental sustainability of an organization's operations. The standard enables organizations to identify and mitigate environmental risks, manage their environmental impacts, and comply with applicable legal and regulatory requirements

EMAS (Eco-Management and Audit Scheme) is another voluntary environmental management standard designed to help organizations improve their environmental performance and become more efficient. It requires organizations to undertake regular environmental audits, evaluate their environmental performance regularly, and implement measures to reduce their environmental impact continually . In India, policies and standards such as the National Action Plan on Climate Change (NAPCC) and National Biodiversity Act (NBA) provide guidance on environmental management in the agricultural sector. However, the adoption of ISO 14001 and EMAS in Indian agriculture is still in its nascent stages. Companies such as ITC Limited, a leading tobacco, and food processing company, have already implemented ISO 14001 in their agricultural operations to promote sustainable farming practices that reduce their environmental impact.

ISO 14001 and EMAS provide several key requirements that organizations must meet to achieve certification. These include the development of an environmental policy statement, identification of significant environmental aspects and impacts, planning and implementation of environmental objectives and targets, establishment of an environmental management program, monitoring and measurement of environmental performance, and regular review and evaluation of the system's effectiveness. In the context of Indian agriculture, meeting these requirements would involve the implementation of sustainable agricultural practices such as integrated pest management, organic farming, conservation of soil and water resources, crop diversification, use of renewable energy, and reduction of waste and pollution. These practices would help reduce the environmental impact of agricultural operations, promote resource conservation, and enhance the long-term sustainability of agriculture in India (Khanna, 2019).

ISO 14001 and EMAS are essential international standards that can be applied to the agricultural sector in India to promote sustainable farming practices that uphold environmental regulations and certification requirements. The adoption of these standards would require the implementation of sustainable practices that conserve natural resources, reduce environmental impact, and enhance the long-term sustainability of Indian agriculture. Organizations must also recognize that environmental management is an ongoing process that requires continual evaluation and improvement to ensure compliance and meet the evolving needs of the environment.

ISO 14001 and EMAS are not only important standards for environmental management in the agricultural sector but are also crucial for enhancing the competitiveness of Indian agriculture in the global market. Farmers and agribusinesses that adopt these standards are better positioned to comply with environmental regulations and customer requirements for sustainable product supply chains. However, the adoption of ISO 14001 and EMAS in Indian agriculture requires increased awareness and understanding of the benefits of these standards. The lack of adequate resources, technical expertise, and institutional capacity in the agricultural sector can also limit their implementation (Chauhan et.al, 2018)

To address these challenges, the government of India has initiated several programs and policies to promote sustainable agriculture practices and enhance the competitiveness of Indian agriculture. These include the Pradhan Mantri Krishi Sinchai Yojana, a national program for sustainable water management, the Soil Health Card Scheme for soil health monitoring, and the Paramparagat Krishi Vikas Yojana, a program for agroecological farming practices. Moreover, the government has also established institutions such as the National Accreditation Board for Certification Bodies (NABCB) and the National Biodiversity Authority (NBA) to promote the adoption of international environmental standards and ensure their effective implementation. The adoption of ISO 14001 and EMAS standards in Indian agriculture is essential for promoting sustainable farming practices, increasing competitiveness, and complying with environmental regulations and market requirements. However, the adoption of these standards requires increased awareness, technical expertise, institutional capacity, and government support. It is imperative that all stakeholders, including farmers, agribusinesses, and policymakers, work together to ensure the effective implementation of these standards to enhance the sustainability and competitiveness of Indian agriculture.

1.4. Best Practices and Case Studies of Companies and Organizations that have Successfully Implemented Environmental Policies and Management Systems in Indian Agriculture

Several companies and organizations have successfully implemented environmental policies and management systems in Indian agriculture to promote sustainable farming practices and enhance the long-term sustainability of the sector. These best practices and case studies provide valuable insights into the benefits and challenges of adopting environmental management standards such as ISO 14001 and EMAS (Chauhan et.al, 2018)

One such company is Mahindra & Mahindra (M&M), a leading Indian manufacturer of tractors and farm equipment. M&M has implemented an environmental management system (EMS) based on ISO 14001 at its manufacturing facility in Nagpur. The company has adopted several sustainable practices such as rainwater harvesting, zero discharge of effluents, and use of renewable energy to reduce its environmental impact. The implementation of the EMS has helped the company reduce its water consumption by 75%, energy consumption by 30%, and CO2 emissions by 63%, while also ensuring compliance with environmental regulations.

Another example is the Sustainable Sugarcane Initiative (SSI) implemented by the Solidaridad Network Asia in partnership with Netafim India. The SSI is a sustainable farming system that promotes practices such as conservation tillage, precise planting, soil fertility management, and weed control to enhance crop productivity, reduce water consumption, and improve soil health. The implementation of the SSI has helped farmers in India increase their yield by up to 47%, reduce water consumption by up to 40%, and reduce greenhouse gas emissions by up to 31%, while also improving their livelihoods.

ITC Limited is another company that has adopted sustainable practices in its agricultural operations and obtained ISO 14001 certification for its environmental management system. The company has implemented practices such as integrated pest management, organic farming, crop diversification, and conservation of soil and water resources to reduce its environmental impact and ensure the sustainable production of its tobacco and food products. The implementation of the EMS has helped the company reduce its water consumption by over 50%, reduce greenhouse gas emissions by over 20%, and increase the use of renewable energy to 56% of its total energy consumption

The Government of India has also implemented several programs and policies to promote sustainable agriculture practices and environmental management in the sector. For example, the Sikkim Organic Mission aims to make Sikkim a fully organic state by promoting practices such as composting, intercropping, and natural pest management. The mission has already achieved 100% organic farming status for over 76,000 hectares of land in the state and has received several awards for its innovative approach to sustainable agriculture.

Moreover, the Maharashtra Agribusiness Network (MABNet) has implemented an online platform that provides farmers with access to information on sustainable farming practices, market trends, and procurement processes. The platform has enabled farmers to improve their productivity, reduce their environmental impact, and increase their income by connecting them with buyers who prefer sustainable and organic products.

Several companies and organizations have successfully implemented environmental policies and management systems in Indian agriculture to promote sustainable farming practices, improve resource efficiency, and comply with environmental regulations. The adoption of practices such as integrated pest management, organic farming, soil fertility management, and conservation of water resources have helped reduce the environmental impact of agriculture operations, improve soil health, and increase productivity. These best practices and case studies provide valuable insights into the benefits and challenges of adopting environmental management standards and highlight the importance of a collaborative approach involving farmers, agribusinesses, and policymakers to promote sustainable agriculture in India.

The Indian Government has implemented several programs and policies to promote sustainable agriculture practices and enhance the long-term sustainability of the sector. One such program is the National Agricultural Higher Education Project (NAHEP), which aims to transform agricultural education and research in India and promote sustainable farming practices. The program provides funding to agricultural universities for the development of new curriculum, research projects, and infrastructure to improve the quality of education and research in the sector. The NAHEP has also implemented the Institutional Development Plan (IDP) to help agricultural universities identify their strengths and weaknesses and develop strategies to address them. The IDP provides a framework for universities to assess their current status, develop a vision for the future, and develop action plans to achieve their goals. The IDP also emphasizes the importance of promoting sustainable agricultural practices and addressing the environmental challenges facing the sector. Several universities have successfully implemented the IDP and adopted sustainable practices in their operations. For example, Dr. Balasaheb Sawant Konkan Krishi Vidyapeeth (DBSKKV) has developed an Integrated Farming System (IFS) for sustainable agriculture. The IFS promotes practices such as crop diversification, intercropping, and organic farming to improve soil fertility, conserve water resources, and reduce the dependence on chemical inputs. The implementation of IFS has helped farmers increase their income, reduce their environmental impact, and improve their resilience to climate change . Another example is the Tamil Nadu Agricultural University (TNAU), which has implemented the IDP to improve its research capabilities and promote sustainable farming practices. TNAU researchers have developed practices such as the System of Millets Intensification (SMI) for enhancing millet productivity, the Nutrient Expert System (NES) for fertilizer management, and the Integrated Pest Management (IPM) for pest control. These practices have helped farmers improve their productivity, reduce their environmental impact, and ensure the long-term sustainability of their operations. The implementation of IDP has also emphasized the importance of promoting sustainable practices in agriculture extension services. For example, Punjab Agricultural University (PAU) has developed the Integrated Farming System Model (IFSM) for extension services. The IFSM promotes practices such as the use of farmyard manure, crop diversification, and conservation agriculture to enhance productivity, conserve natural resources, and reduce the environmental impact of agriculture. The implementation of IFSM has helped farmers increase their income, reduce their input costs, and improve their resilience to natural disasters.

The Indian Government's programs and policies such as NAHEP and IDP are promoting sustainable agriculture practices and enhancing the long-term sustainability of the sector. The implementation of IDP has led to the adoption of several sustainable practices by agricultural universities, which have helped farmers improve their productivity, reduce their environmental impact, and ensure the long-term sustainability of their operations. These initiatives not only promote sustainable agriculture practices but also improve the livelihoods of farmers and enhance their resilience to climate change.

1.5. The Challenges Faced in Implementing and Adhering to Environmental Policies and Management Systems in Indian Agriculture

Implementing and adhering to environmental policies and management systems in Indian agriculture presents several challenges. These challenges include inadequate awareness about the benefits of sustainable practices, lack of economic incentives, weak institutional frameworks, and limited access to finance and technology. Addressing these challenges requires a collaborative approach from various stakeholders, including the government, farmers, and the private sector. This paper proposes solutions to overcome the challenges faced in implementing and adhering to environmental policies and management systems in Indian agriculture.

Inadequate awareness about the benefits of sustainable practices is one of the significant challenges hindering the implementation of environmental policies and management systems in Indian agriculture. Many farmers are not aware of the long-term benefits of adopting sustainable practices, including improved soil fertility and reduced environmental impact. Therefore, awareness-raising campaigns, capacity-building, and training programs can help farmers understand the importance of sustainable practices. Agricultural extension services can also promote sustainable farming practices to farmers and help them access information related to environmental policies and management systems. A lack of economic incentives is another challenge in promoting sustainable agriculture practices. Currently, conventional farming is often more profitable than sustainable farming practices, such as organic or agroforestry practices. To encourage the adoption of sustainable practices, governments can establish economic incentives such as subsidies for eco-friendly inputs, tax credits, or price premiums for sustainable products. These incentives can motivate farmers to adopt sustainable practices by making them financially feasible.

Weak institutional frameworks, such as insufficient implementation of regulations and inadequate infrastructure, pose a significant challenge to implementing environmental policies and management systems. To address this challenge, institutions must be strengthened to ensure compliance with environmental regulations. Governments can improve institutional capacity by allocating more resources towards strengthening regulatory agencies and organizations that promote sustainable agriculture practices. The government can also invest in infrastructure, such as irrigation systems and storage facilities, to ensure the smooth implementation of agricultural regulations.

Limited access to finance and technology is another significant challenge that hinders the adoption of sustainable agricultural practices. Many farmers lack access to affordable finance to purchase inputs and invest in sustainable technologies. Governments can address this challenge by establishing appropriate credit facilities, such as low or zero-interest loans, to help farmers adopt sustainable farming practices. Moreover, technology transfer programs can be implemented to help farmers access modern technologies that promote sustainable agricultural practices.

Finally, the participation of the private sector is essential in ensuring adherence to environmental policies and management systems. The private sector can offer environmental audit services to help farmers comply with regulations, develop eco-friendly products, and establish environmental standards that promote sustainable agriculture practices. The private sector can also provide technological support and finance to promote the adoption of sustainable practices, particularly in high-value agricultural products.

In conclusion, implementing and adhering to environmental policies and management systems in Indian agriculture requires addressing several challenges, including inadequate awareness about the benefits of sustainable practices, lack of economic incentives, weak institutional frameworks, and limited access to finance and technology. To overcome these challenges, a collaborative approach from various stakeholders, including the government, farmers, and private sector, is necessary. Addressing these challenges will require policy interventions such as promoting awareness-raising campaigns, offering economic incentives, strengthening institutional frameworks, providing access to finance and technology, and encouraging private sector participation.

Appendix

Designing Effective Environmental Policies and Management Systems for Indian Agriculture

Introduction

The Indian agriculture industry has been facing significant challenges related to sustainable development, including the impact of climate change, pollution, and resource depletion. To address these challenges, it is crucial for companies and organizations in the Indian agriculture industry to implement effective environmental policies and management systems. This sub-chapter aims to design a proposal for a company or organization in the Indian agriculture industry detailing how they can implement effective environmental policies and management systems for sustainable development. An implementation plan with potential challenges and solutions will also be presented.

Proposal

The proposed company/organization is a medium-scale farming enterprise that grows cash crops such as cotton, soybean, and maize. The enterprise has 50 acres of agricultural land and employs around 100 workers. The goal of this proposal is to create a comprehensive environmental policy that promotes sustainability in the farming practices of the enterprise.

Step 1: Formulate Environmental Policy

The Environmental Policy forms the foundation for the implementation of environmental policies and management systems. It is the first and most critical step of the proposal and involves the creation of a policy statement which outlines the enterprise's commitment to sustainability. The policy statement should be followed by the adoption of key principles for sustainable development, including reduction of environmental footprint, efficient use of resources, and adherence to environmental regulations and standards.

Step 2: Identification of key areas of impact

The next step is to conduct an environmental impact assessment to identify the areas where the enterprise's activities have the most significant impact. The impact assessment would help the enterprise to identify the areas in which they need to implement sustainable practices and promote sustainable development.

Step 3: Implementation of Sustainable Practices

The identified key areas of impact will be the focus of the implementation of sustainable practices. The practices to be implemented are those that have the most significant positive impact on the environment. Some recommended sustainable practices to the enterprise would include:

1. The use of organic fertilizers and pesticides in agricultural production.
2. Implementation of good water management practices, including rainwater harvesting and efficient irrigation systems.
3. The use of renewable energy sources like solar panels and wind turbines to reduce fossil fuel usage and emissions.
4. Promotion of biodiversity conservation measures such as the reduction of monoculture and cultivation of cover crops.

Step 4: Identify Key Performance Indicators

To evaluate the success and impact of the proposed environmental policies and management systems, the enterprise will need to identify relevant key performance indicators (KPIs) and monitor them over time. KPIs could include waste reduction, energy efficiency, water conservation, and GHG emissions reduction.

Potential Challenges and Solutions

There may be some challenges associated with the implementation of environmental policies and management systems in the Indian agriculture industry, which are unique to the sector. Some of these challenges include:

1. Resistance to change from traditional practices amongst the workforce.

   Solution: The enterprise can involve the workers in the decision-making process and educate them on the benefits of sustainability practices.

2. High initial capital investment required for renewable energy and advanced technology implementation.

   Solution: The enterprise can explore alternative financing options like government subsidies and loans.

3. Limited availability of cost-effective organic fertilizers and pesticides.

   Solution: The enterprise can explore collaboration opportunities with companies that produce or distribute organic inputs.

The implementation of effective environmental policies and management systems aligns with the Indian government's objectives of promoting sustainable development in agriculture. This proposal outlines the steps that an agricultural enterprise can take to implement effective environmental policies and management systems for sustainable development. The identified potential challenges and solutions can aid the enterprise in devising a plan that is adapted to their needs, overcoming potential obstacles, and ensuring the long-term success of the proposal.

Chapter Section Summaries

1-1: The current state of environmental policies in Indian agriculture and the challenges faced in implementation. This section provides an overview of the state of environmental policies in Indian agriculture and the challenges faced in implementing and adhering to them. It highlights the factors contributing to the challenges and how they impact sustainable agricultural development.

1-2: The importance of implementing environmental policies and management systems in Indian agriculture to ensure sustainable development. In this section, the importance and benefits of implementing environmental policies and management systems in Indian agriculture are discussed. The section also highlights the role of sustainable agricultural development in ensuring food security and the well-being of farmers and rural communities.

1-3: International standards and their application to Indian agriculture. This section discusses international standards, including ISO 14001 and EMAS, and their application in the context of Indian agriculture. It highlights the benefits of complying with international standards and how they can help companies and organizations achieve environmental sustainability.

1-4: Best practices and case studies of successful implementation of environmental policies and management systems in Indian agriculture. This section explores the best practices and case studies of companies and organizations that have successfully implemented environmental policies and management systems in Indian agriculture. It analyzes the strategies that contributed to their success and draws insights and lessons that can be learned from them.

1-5: Designing and proposing solutions to overcome challenges in implementing and adhering to environmental policies and management systems in Indian agriculture. In this section, potential solutions to the challenges discussed in Chapter 1 are explored, including awareness-raising campaigns, economic incentives, strengthening institutional frameworks, technology transfer programs, and private sector participation. It also discusses the need for a collaborative approach involving multiple stakeholders to ensure successful implementation and adherence to environmental policies and management systems in Indian agriculture.

Discussion Questions

1. What are some of the key challenges faced in implementing environmental policies in the Indian agriculture industry? How can these challenges be addressed?
2. Explain the importance of integrating environmental policies and management systems in Indian agriculture to ensure sustainable development. Provide examples of how this can be achieved in practice.
3. Evaluate the role of international standards such as ISO 14001 and EMAS in the context of Indian agriculture. What are some of the key requirements necessary for compliance with these standards in India?
4. Compare and contrast case studies of companies and organizations that have successfully implemented environmental policies and management systems in Indian agriculture. What key insights can be drawn from these examples?
5. Design a proposal for a company or organization in the Indian agriculture industry detailing how they can implement effective environmental policies and management systems for sustainable development. Provide an implementation plan and identify potential challenges and solutions.

Application Case Studies

Case Study 1: Agri-Export Zone (AEZ) scheme

The AEZ scheme is a significant initiative launched by the Indian government to promote sustainable agriculture practices, increase agricultural exports, and improve the livelihoods of farmers. The implementation of this scheme, however, presents several challenges related to environmental sustainability, social inclusiveness, and governance. It requires a coordinated effort from various stakeholders, including the government, farmers, and private sector to ensure its success and long-term sustainability. As such, it is essential to reflect on the policies and management systems in place that are conducive to promoting sustainable agriculture practices.

Imagine a scenario where a group of local farmers in India is struggling to sell their produce due to lack of market access, low productivity, and poor quality. They are facing severe financial difficulties, and this is impacting their ability to sustain their agricultural operations. The government has launched the Agri-Export Zone (AEZ) scheme to improve their situation by promoting the export of agricultural products and increasing the revenue for farmers. However, the implementation of this scheme is challenging, and it requires farmers to adopt sustainable agricultural practices, including good governance and social inclusiveness.

Reflective Questions

1. How can the government ensure that the AEZ scheme is implemented in a socially inclusive manner that benefits all farmers, regardless of their socio-economic status?
2. What measures can be taken to promote environmental sustainability in Indian agriculture and ensure that the AEZ scheme aligns with the country's environmental policies and management systems?
3. How can the AEZ scheme be made more efficient and effective in increasing agricultural productivity and revenue? What metrics should be used to evaluate its impact?
4. What role can technology play in implementing the AEZ scheme and enhancing sustainable agriculture practices in India? How can farmers be encouraged to adopt technological innovations?
5. What are the potential challenges and opportunities in implementing the AEZ scheme in a diverse country like India, with varying levels of agricultural development, infrastructure, and resources? How can these challenges be addressed?

The reflective questions above provide thought-provoking prompts to consider in developing solutions to address the environmental challenges facing India's agriculture industry.

Case Study 2: National Seeds Corporation (NSC)

The National Seeds Corporation (NSC) is a significant enterprise in India responsible for producing and distributing certified seeds that improve the quality and availability of seeds for farmers. However, the corporation must contend with environmental sustainability, social inclusion, and good governance challenges in its operations. The questions raised below provide an opportunity to consider how environmental policies and management systems can be integrated into its operations to ensure that the NSC meets its sustainability goals. It is important to reflect on these issues to ensure that the NSC operates in an environmentally and socially responsible manner that benefits local communities and the agriculture industry.

Imagine a scenario where a group of farmers in India is struggling to produce high-quality crops due to the use of substandard or fake seeds. They are facing severe financial difficulties, and the crops they grow are not suitable for market consumption. The government has established the National Seeds Corporation (NSC) to improve the quality and availability of seeds for farmers. However, the NSC faces several challenges in ensuring environmental sustainability, social responsibility, and good governance in its processes and operations.

Reflective Questions

1. How can the NSC ensure that the seeds it produces are suitable for the environmental conditions and soil composition of the local communities where they will be used? What measures should be implemented to promote environmentally sustainable seed production and distribution?
2. What role can local communities play in seed production and distribution, and how can the NSC involve them in its operations? How can the NSC ensure that local communities benefit equitably from the seed production and distribution system?
3. How can the NSC ensure that its operations comply with relevant environmental policies and management systems? What are the potential challenges in adhering to these policies and systems, and how can these challenges be addressed?
4. What measures can be taken to promote greater transparency and accountability in the NSC's seed production and distribution system? What metrics should be used to evaluate the corporation's social and environmental impact?
5. What role can technological innovations play in enhancing the sustainability and efficiency of the NSC's seed production and distribution system? How can farmers be encouraged to adopt technological innovations in seed production and distribution?

The reflective questions above provide thought-provoking prompts to consider in developing solutions to address the environmental challenges facing India's agriculture industry.

References

AgriInputs. (2020, August 18). Challenges Faced by Indian Agriculture and Possible Solutions. Retrieved September 28, 2021, from https://www.agriinputs.com/blog/challenges-faced-by-indian-agriculture-and-possible-solutions/

Bhatia, Aarti, et al. "Sustainable Agriculture in India: An Overview." International Journal of Agriculture, Environment and Biotechnology, vol. 9, no. 2, 2016, pp. 337-346, https://

Chauhan, A., & Singh, A. K. (2018). Issues and prospects for adoption of environmental management system in Indian agriculture. Journal of Cleaner Production, 183, 253-267.

Chauhan, A., & Singh, A. K. (2018). Issues and prospects for adoption of environmental management system in Indian agriculture. Journal of Cleaner Production, 183, 253-267.

Das, S., & Hati, K. M. (2019). Sustainable Agriculture Practices and Policies for Climate Change Mitigation and Adaptation in India. Journal of Agricultural Science and Technology, 4(3), 47-59.

Dr. Balasaheb Sawant Konkan Krishi Vidyapeeth. (n.d). Integrated Farming System. Retrieved from https://dbskkv.org/Extension/IntegratedFarmingSystem.php

E-National Agriculture Market (e-NAM). (n.d.). Retrieved September 22, 2021, from https://nemmp.gov.in/enam-dashboard

European Commission. (2020). Eco-Management and Audit Scheme (EMAS). Retrieved from https://ec.europa.eu/environment/emas/about-emas/index_en.htm

Farmers' Produce Trade and Commerce (Promotion and Facilitation) Act, 2020 - https://www.prsindia.org/sites/default/files/bill_files/Farmers%20Produce%20Trade%20and%20Commerce%20%28Promotion%20and%20Facilitation%29%20Act%2C%202020.pdf

Food and Agriculture Organization. (2018). International agricultural trade and standards. Retrieved from http://www.fao.org/3/ca6410en/ca6410en.pdf

Food and Agriculture Organization of the United Nations. (2018). Sustainable Agriculture and Rural Development. Retrieved September 28, 2021, from http://www.fao.org/3/i9912en/I9912EN.pdf

Government of Sikkim. (2016). Sikkim Organic Mission. Retrieved from https://sikkim.gov.in/state-initiatives/sikkim-organic-mission

Indian Council of Agricultural Research. (2021). Institutional Development Plan. Retrieved from https://nahep.icar.gov.in/idp

Indian Council of Agricultural Research. (2021). National Agricultural Higher Education Project. Retrieved from https://nahep.icar.gov.in/

International Organization for Standardization. (2015). ISO 14001 - Environmental Management. Retrieved from https://www.iso.org/iso-14001-environmental-management.html

International Organization for Standardization. (2015). ISO 14001 - Environmental Management Requirements. Retrieved from https://www.iso.org/standard/60857.html

ITC Limited. (2021). Sustainable Agriculture Practices. Retrieved from https://www.itcportal.com/sustainability/sustainability-initiatives/sustainable-agriculture-practices.aspx

Jat, H. S., Singh, R. K., Singh, Y., Singh, M., & Ravinder, K. (2017). Challenges and approaches for sustainable management of dryland agriculture: A case study from India. Frontiers of Agricultural Science and Engineering, 4(2), 127-136. https://doi.org/10.15302/J- FASE-2017076

Khanna, R., & Sahoo, A. K. (2019). Sustainable Agriculture in India: Challenges, Issues, and the Way Forward. Current Agriculture Research Journal, 7(2), 233-238.

Kumar, P., Ganguly, D., & Singh, A. P. (2020). Environmental impact of Indian agriculture: A review of the current state and future prospects. Global Journal of Environmental Science and Management, 6, 107-126.

Mahapatra, I. C., Pal, S., Tripathi, A. K., & Chanakya, H. N. (2019). Impacts of agricultural practices on soil and water quality in India: A review. Land, 8(8), 112.

Maharashtra Agribusiness Network. (n.d.). MABNet. Retrieved from http://mabnet.in/ Mahindra & Mahindra. (2019). Environmental Stewardship. Retrieved from https://www. mahindra.com/about-us/sustainability/environmental-stewardship

Ministry of Agriculture & Farmers Welfare. (2019). Agricultural Statistics at a Glance 2018. New Delhi: Government of India.

Ministry of Agriculture and Farmers Welfare. (2021). Schemes. Retrieved from https://www.agriculture.gov.in/en/schemes

Ministry of Environment, Forest, and Climate Change. (2017). National Action Plan on Climate Change. Retrieved from http://www.moef.gov.in/sites/default/files/nat%20action%20plan%20on%20climate%20change.pdf

Ministry of Environment, Forests & Climate Change. (2015). India's Intended Nationally Determined Contribution: Working towards Climate Justice. New Delhi: Government of India.

Ministry of Environment, Forests, and Climate Change. (2020). Environmental Standards and Guidelines. Retrieved from http://www.moef.gov.in/environmental-standards-and-guidelines

Moody, A. (2015). Environmental policies and agricultural sustainability in India. Environmental Economics, 6(1), 45-54. https://doi.org/10.21511/ee.06(1).2015.05

National Action Plan on Climate Change. Ministry of Environment, Forest and Climate Change, Government of India, 2008, http://moef.gov.in/wp-content/uploads/2017/06/ national_action_plan_on_climate_change.pdf. Accessed 12 Aug. 2021.

National Agricultural Higher Education Project (NAHEP). (n.d.). Retrieved September 22, 2021, from https://nahep.icar.gov.in/

National Agricultural Technology Project (NATP). (n.d.). Retrieved September 22, 2021, from https://nrdp.agritech.tnau.ac.in/projects/national-agricultural-technology-project-natp/

National Mission for Sustainable Agriculture. Department of Agriculture, Cooperation and Farmers Welfare, Government of India, https://nmsa.dac.gov.in/. Accessed 12 Aug. 2021.

National Mission on Sustainable Agriculture (NMSA). (n.d.). Retrieved September 22, 2021, from http://nmsa.dac.gov.in/

National Program for Organic Production (NPOP). (n.d.). Retrieved September 22, 2021, from https://www.apeda.gov.in/

Pal, Subrata, and Amit Banerjee. "Dilemma of Indian Agriculture: Environmental Issues and Policies." Journal of Agricultural and Environmental Ethics, vol. 33, no. 3, 2020, pp. 413-422, https://doi.org/10.1007/s10806-020-09806-6. Accessed 12 Aug. 2021.

Paramparagat Krishi Vikas Yojana. Ministry of Agriculture and Farmers Welfare, Government of India, https://pkkv.csccloud.in/. Accessed 12 Aug. 2021.

Pradhan Mantri Fasal Bima Yojana (PMFBY) - https://pmfby.gov.in/

Precision farming - https://www.agriculture.gov.in/policyresolution.aspx?id=6&subid=10

Punjab Agricultural University. (n.d.). Integrated Farming System Model. Retrieved from https://www.pau.edu/content/pdf/success/IFSM.pdf

Rainfed Area Development Program (RADP). (n.d.). Retrieved September 22, 2021, from http://www.dacnet.nic.in/radp/radp.htm

Singh, R. K., and Kumar, A. (2017). Sustainable agriculture and challenges in Indian perspective. International Journal of Agriculture Sciences, 9(37), 4243-4247.

Singh, R.K., et al. "Sustainable Agriculture in India: Current Status and Future Directions." Proceedings of the Indian National Science Academy, vol. 86, no. 2, 2020, pp. 243-255, https://link.springer.com/article/10.1007%2Fs40011-019-01104-w. Accessed 12 Aug. 2021.

Soil degradation and sustainable agriculture - https://www.researchgate.net/publication/289596675_Soil_Degradation_and_Sustainable_Agriculture_in_India

Soil Health Card Scheme. Department of Agriculture, Cooperation and Farmers Welfare, Government of India, https://soilhealth.dac.gov.in/. Accessed 12 Aug. 2021.

Solidaridad. (n.d.). Sustainable Sugarcane Initiative. Retrieved from https://www.solidaridadnetwork.org/what-we-do/sustainable-sugarcane-initiative

Tamil Nadu Agricultural University. (n.d.). Institutional Development Plan. Retrieved from https://tnau.ac.in/idp/

Tilak, K. V. B. R., & Chand, R. (2016). Environment regulations and Indian agriculture: Issues, challenges and the way forward. Journal of Applied and Natural Science, 8(4), 2461-2465. https://doi.org/10.31018/jans.v8i4.978

United Nations Industrial Development Organization. (2013). Introduction to Environmental Management Systems (EMS). Retrieved from https://www.unido.org/sites/default/files/2013-03/03-00264%20EMS%20brochure_0.pdf

Verma, Raj Kumar, et al. "Sustainable Agriculture in India: Concept, Status, and Challenges." Indian Journal of Agricultural Sciences, vol. 88, no. 11, 2018, pp. 1689-1706, https://www.researchgate.net/publication/326739180_Sustainable_Agriculture_in_India_Concept_Status_and_Challenges. Accessed 12 Aug. 2021.

Water management in Indian agriculture - https://www.indiawaterportal.org/article/challenges-and-opportunities-water-management-indian-agriculture
www.mendeley.com/catalogue/2a8182ef-2b6e-3378-a1a4-b4ade5e40acf/. Accessed 12 Aug. 2021.

2

Social Policies and Management Systems & Processes in Indian Agriculture Context

Dur Khan

Assistant Professor, JBIMS

Learning Objectives

2.1. To define the current state of Indian agriculture and the challenges faced

2.2. To understand the current state of social policies in Indian agriculture

2.3. To evaluate the importance of implementing management systems in Indian agriculture

2.4. To analyze the significance of key international standards such as ISO 26000 and SA8000 in the context of Indian agriculture

2.5. To appraise the best practices and case studies of successful implementationof social policies and management systems in Indian agriculture

2.5. To discuss the way forward for the policymakers

1.1. Overview of the Indian Agriculture Sector and Challenges Faced

According to the Indian economic survey of 2018, more than 50% of the Indian workforce is employed in the agricultural sector. Agriculture contributed to 17–18% to the country's GDP. Agriculture's share in India's economy is less compared to the industrial and services sectors. The industrial and service sectors have higher growth rates. The significance of the agricultural sector in India's economic and social growth goes well beyond. Nearly 60% of India's families depend on rural incomes. In addition to this, the majority of India's poor population (around 70 %) lives in rural areas. Also, as the population is growing at a fast rate, food security in India depends on producing cereal crops and in increasing the production of fruits, vegetables and milk to sustain the needs of a growing population. To achieve this we need a productive, diversified, competitive, and sustainable agricultural sector. India is the world's largest producer of pulses, spices and milk. It has the world's largest cattle herd and the largest area under wheat, rice, and cotton. India is also the second largest producer of rice, wheat, cotton, sugarcane, farmed fish, sheep & goat meat, fruit, vegetables and tea.

The Indian agricultural sector has undergone immense change post the Green Revolution in the late 1960s. The entire agriculture strategy was focused on a single motto of 'grow more food'. Because of the green revolution, the production of food has increased, fertility of soil improved. India as a nation became self-sufficient in food production and the because of surplus production the export also increased. Modern irrigation methods like drip irrigation, more efficient fertilizers, stronger and more resistant pesticides, and newly developed seeds aided in proficient growth of crops. India experienced drastic increases in crop production as a result of such innovative developments in agriculture.

However, there are alarms concerning the sustainability of current production approaches. Heavy use of chemical fertilizers and insecticides, expansion of irrigation, and crop specializations favouring a few selected crops which were the main sources of growth of agriculture post-Green Revolution, all this created destruction of natural resources and the environment. Heavy subsidy and free supply of power for irrigation led to irresponsible, indiscriminate, and overuse of water. It also brought serious changes in crop choices. In spite of excess production of cereals and a multi-fold increase in the availability of vegetables, fruits, and milk for domestic use, a large population still lives in hunger and many remain undernourished.

Added concern is a farmer's income. There is still a huge inequality between non-agricultural and agricultural income. The small and decreasing land size, surplus labour force, low efficiency and poorly working markets are the key reasons for low per-farmer income in the country. If the current methods and policies are not enhanced, there is less chance to reduce this wage gap in the near future. Also, apart from concentrating on raising income from farming, there is a need for creating avenues for non-agricultural income. The agricultural households need to earn income from non-farm sources.

Agriculture growth and social policies are meaningfully helpful for reducing poverty and increasing the per capita income of the low-income classes in rural India. Above and beyond comprehensive growth, agricultural developments are important for healthiness and nourishment, sustainability, climate change, and the quality of life in the country. The present government has set a vision for a new India that involves 'Sabka Saath Sabka Vikas'. Transformation of the agriculture sector is crucial for attaining this dream, as 45% of the labour force in the country is working in the agriculture sector (National Statistical Office 2021a).

Strategies for the 21st Century must concentrate on efficiency in farm practices, research and development, innovation and employment generation. The approach needs to shift to efficient growth which requires the upgradation of agricultural technology, application of contemporary skills in farm practices, new inventions in agriculture, and lowering wastage in use of fertilizer, water and other inputs. Digital technology can also play an important role in improving efficiency by easily disseminating knowledge and technology to farmers.

Increasing cost of production related to use of current technology is resulting in the consideration of alternative methods of farming. Methods rooted in traditional Indian methods of farming, like natural farming and organic farming, are proposed in place of chemical-based farming with the purpose to cut cost of production. It will also reduce the adverse effects of chemicals on health, environment and natural resources.

Improving the agricultural efficiency is determined by robust and vibrant Research & development (R&D) by public or private sector. Public sector R&D in the country faces challenges like resource constraints, disciplinary fragmentations, and lack of drive and motivation. On the other hand, fascinating innovation and changes in agriculture are seen in advanced institutions in private sector. Private sector investment in agriculture-based R&D in India is also low due to the complex nature of Intellectual Property Right (IPR) regime in the country. Therefore, the gap between national and worldwide agricultural innovations is rising, and many captivating changes experienced in worldwide agriculture are absent in the country. In the absence of domestic R&D achieving global standard, agriculture is losing its competitive advantage. There is a need to enable easy access to our farmers to global technology, high quality seeds and other knowledge products.

Recent growth in technologies like automation, artificial intelligence (AI), big data, internet of things (IOT) and machine learning are also limiting the capacity of the non-agriculture sector to absorb the workers from agriculture. This has put renewed attention on agriculture to generate useful employment opportunities in the value-added activities of the sector.

In era of capital-intensive production favoured by the manufacturing sector and expected threats to jobs by developing technological innovations, there is a need to reconsider the strategy of shifting workers from agriculture to manufacturing and services. India should discover opportunities for generating blue-collar jobs in and around agriculture. This also looks desirable, as there is a grave deficiency of skilled workforces in agriculture.

A vital cause for the sluggish shift of farm youths to the non-agriculture sector is their strong preference for middle-income jobs. But, in many cases, the mandatory skill and capability for such jobs are absent. Medium, small, and micro enterprises (MSME) are labour-intensive and they seem to be a suitable substitute for rural employment generation. Linking agri-food processing to production over efficient value chains, contract farming and direct linkage between factory and farm offer significant opportunities for rural employment generation as well as for raising farmers' income.

The motives for workers choosing to move from agriculture to non-agriculture are relatively small wages in agriculture, manual work, and uneven employment. These three difficulties can be addressed by using advanced methods in agriculture and postharvest activities. These can be harnessed by evolving and promoting new farm models centred on the knowledge- and skill-based agriculture and the post-harvest value addition at the farm itself. Pradhan Mantri Kaushal Vikas Yojana (PMKVY) can play a major role in this by promoting and imparting skills essential in modern agriculture.

The performance of the agriculture and the allied sector has been resilient to the wave of COVID-19 pandemic. The sector grew at 3.6 percent in 2020-21 and improved to 3.9 percent in 2021-22. Still, the fragmentation of landholdings has led to substitute sources such as livestock, fishery and wage labour becoming meaningfully important for an agricultural household. The growing importance of allied sectors including animal husbandry dairying and fisheries in the growth and income of the farmers specifies that the focus needs to shift more towards harnessing the potential of allied activities. There is also a need to expand the efficiency of small and marginal farmers through the development and use of small-holding farm technologies.

Crop diversification to oilseeds, pulses, and horticulture needs to be prioritized by addressing central problems like irrigation, investment, credit, and markets. The government has adopted the use of Minimum Support Prices as a signal to encourage crop diversification. Synchronized acts from the State Governments will enable the shift of the sector to high-value and less resource-consuming crops to enable the realization of the objective of doubling farmers' income in a justifiable way.

1.2. Overview of the Current State of Social Policies in Indian Agriculture

To understand the progress and impact of social policies in Indian agriculture, we must first define the term "social policy". Social policies take into consideration the ways societies across the world address needs for security, education, work, well-being, and health. The protective policies fundamentally give safety for employment, income, well-being, etc. On the other hand, the productive policies expand the productive capacity of the population, this includes education, training, and healthcare. The term social policy has a very broad and indefinable definition, it brings under its horizon a widespread range of government policies affecting numerous facets of human life. These social policies in India are planned and implemented at the Centre and State levels. Some Government Policies and Interventions include;

1. Pradhan Mantri Fasal Bima Yojana (PMFBY)

Agriculture plays a significant role in India's economy. Census 2011 reported that around 54.6% of the total workforce of India is engaged in agricultural and related sector activities. Looking at the importance of the agriculture sector, the Government of India has taken several steps for its expansion in a sustainable fashion. There are steps being taken to improve the income of farmers. To mitigate the risk in the agriculture sector, the scheme "Pradhan Mantri Fasal Bima Yojana" (PMFBY) was launched in 2016. It is an insurance service for farmers for their crop production. PMFBY intends to provide a comprehensive insurance cover against crop failure. This will help in stabilizing the farmers' income.

2. Doubling of Farmers' Income (DFI)

The government constituted a Committee in April 2016 to inspect issues relating to "Doubling of Farmers Income" and recommend approaches to accomplish the same. A report was submitted to the Government by the committee in September 2018. Later a Sanctioned Body was set up in 2019 to monitor and review the progress as per the recommendations of the committee. The Committee has recognized seven sources of revenue growth for farmers namely improvement in crop productivity; progress in livestock productivity; resource use efficiency or savings in the cost of production; surge in the cropping intensity; diversification towards high value crops; advancement in real prices received by farmers; and shift from farm to non-farm jobs.

3. Atma Nirbhar Bharat – Agriculture

Atmanirbhar Bharat Abhiyaan (ABA) means and signifies Self-Sufficient India. It is the vision of a new India announced by the Hon'ble Prime Minister Shri Narendra Modi in the year 2020. ABA aims to strengthen Infrastructure, Logistics, Capacity Building, Governance and Administrative Reforms in agricultural sector. The components of the Atma Nirbhar Bharat – Agriculture include Agriculture Infrastructure Fund (AIF) worth Rs. 1 lakh crores including Rs 500 crores for The National Beekeeping and Honey Mission (NBHM). In addition to this some market reforms have also been rolled out which include;

- Model APLMC Agricultural Produce and Livestock Market Committee (Promotion & Facilitation) Act, 2017
- Establishment of 22,000 Gramin Agriculture Markets (GrAMs) as aggregation platforms
- The Farmers (Empowerment & Protection) Agreement on Price Assurance and Farm Services Act, 2020
- Agri-Export Policy, which targets to increase the agri-exports
- The Farmers Produce Trade and Commerce (Promotion & Facilitation) Act, 2020
- Amendments to Essential Commodities Act, 1955

The Government also created Corpus Funds for the benefit of farmers such as Micro Irrigation Fund of Rs. 5,000 crores and Agri-Marketing Fund to strengthen National Agriculture Market (eNAM) and Agricultural Produce and Livestock Market Committee (GrAMs) of Rs. 2,000 crores

4. The Farmers Produce Trade and Commerce (Promotion and Facilitation) Act 2020

According to this act, State governments are prohibited from charging any market fee, cess outside APMC areas. This will allow intra-state and inter-state trade of farmers' produce, providing more options to farmer for sale of their farming produce. This Act is a very vital step to provide a more competitive and hassle-free eco-system where farmers and traders have the choice to sell their produce in an efficient, transparent and competitive environment and to get remunerative prices.

"The Farmers' Produce Trade and Commerce (Promotion and Facilitation) Act, 2020" does not substitute the State APMC Act. This act will motivate APMCs to improve their efficiency and provide better facilities to farmers.

This will act as an additional channel while the existing APMCs already facilitating farmers in obtaining better remunerative prices for their produce. The provisions of The Farmers' Produce Trade and Commerce (Promotion and Facilitation) Act, 2020" aims to yield compensable prices to the farmers by successful enforcement of improved marketing practices.

5. The Farmers' (Empowerment and Protection) Agreement on Price Assurance and Farm Services Act, 2020

Through this act, the Government of India provides a countrywide framework on farming agreements. It protects farmers and allows them to engage with agribusiness firms, processors, wholesalers, exporters or retailers to sell the future farming produce at a mutually agreed price framework in a fair and transparent way. The objective is to make agriculture less risky at various stages.

These agreements enable scaling of investment by industry for production and processing of high value agriculture produces. The farmers also have a benefit of operational efficiency. The framework mitigates the risks for the farmers and enhance their income. The act is instrumental in creating effective and conducive policy regime for agreements by laying down of procedures and systems. This will enable creation of an eco-system for the holistic development of the agriculture sector.

6. Kisan Credit Card (KCC)

The idea behind KCC is to provide easy concessional institutional agri-credit to all the farmers for meeting the short-term requirements like buying of seeds and fertilizers, raising and marketing of crops, working capital and production needs. Kisan Credit Card scheme is implemented by Commercial Banks, Cooperative Banks and Rural Regional Banks (RRBs).

To cover all eligible farmers under KCC in a mission mode, KCC saturation drive was launched in 2019. To facilitate easy access to KCC, processing fees, inspection, ledger folio charges and all other service charges for fresh issue/ renewal of KCC were waived off, a time limit of 14 days was fixed for issuing KCC after receipt of application and simplified application form were made available.

The farmers engaged in activities relating to Animal Husbandry and Fisheries have been included in the KCC scheme. In collaboration with Department of Financial Services (DFS) a second phase of KCC saturation drive was initiated by Department of Agriculture Cooperation & Farmers Welfare (DAC&FW)

in mid-February, 2020 with the goal to create universal access to institutional concessional credit to all farmers.

7. PM-KISAN

PM Kisan is a scheme of Central Government with 100% funding from Government of India. Operational from 2018, this scheme provides income support of 6,000/- per year in three equal instalments to families of small and marginal farmer having combined land holding of up to 2 hectares of land. Definition of family for the scheme includes husband, wife and their minor children. As per scheme guidelines, the State Government and UT administration identifies the farmer families which are eligible for support. The fund gets directly transferred to the bank accounts of the beneficiaries. There are some Exclusion Categories also in the scheme.

8. Farmer Producer Organization (FPO)

FPO means and includes Farmer-Producers' Organization incorporated or registered either under Part IXA of Companies Act or under Co-operative Societies Act of the concerned States and designed for the purpose of creating economies of scale in production and marketing for agricultural and related sector. These FPOs must be independent and shall be protected from all types of interference including political interference in election process and in day today management using suitable provisioning in their Memorandum of Association and Bye-laws with a view to inspire healthy development of FPO.

9. Commission for Agricultural Costs and Prices

Commission for Agricultural Costs and Prices (CACP) is set up with a view to develop a stable and integrated price structure for the crops. It is mandated to advice on the price policy (Minimum Support Price) of 23 crops. These include Fair and Remunerative prices (FRP) of seven cereal crops (paddy, wheat, jowar, bajra, maize, ragi and barley), five pulse crops (gram, tur, moong, urad and lentil), seven oilseeds (groundnut, sunflower seed, soybean, rapeseed-mustard, safflower, niger seed and sesamum), copra/dried coconut, cotton, raw jute and sugarcane.

In the Price Policy Reports, CACP submits recommendations to the government each year, separately for five groups of commodities (Kharif crops, Rabi crops, Sugarcane, Raw Jute and Copra). The Commission seeks views of various Central Ministries, State Governments, Farmers, Farmers Association, Research Institutes and other stakeholders before preparing these five price policy reports.

1.3. Overview of Management Systems & Processes in the Indian Agriculture Context

Agricultural management includes the discipline of food production. It deals with agricultural methods, the domestication of animals, and the processing of food. There are many farming jobs that necessitate management, particularly on projects in the scientific disciplines and with farm labour. With the varying dynamics in the global food production systems, this field is one of the fastest-growing filed in the world. Ideas like productivity and safety play an important role in the food supply of the world, thus agricultural management has high potential. The goal today is to endeavour for innovative ways to produce the essential quantity of food for the rising population while sustaining the quality that society expects from its food sources.

Agriculture Management is a strategy in management for organizing and implementing agricultural decisions. It involves livestock, crops, aquaculture, agroforestry, etc., to make the most of output and profitability. It allocates resources, develops strategies, and plans operations to warrant profitable, sustainable, and productive farming. Agriculture management is concerned with the organization of resources, planning their use, and controlling of plans both during and after their implementation.

Agriculture Management can be in-depth and expensive. The increasing demand for food production to feed the rising population has amplified the pressure among farmers. The execution of smart farm management tactics will enhance productivity and drive maximum profits for agri-businesses.

Agriculture management provides a seamless system where all activities of farming are well organized along with its optimum utilization of resources for less cost and bigger profits. Agriculture management practices include planning, organizing, implementing, and monitoring agricultural activities like Irrigation, crop rotation, soil management, harvesting, contour ploughing, tillage, post-harvest management, organic farming and sustainable agricultural practices among others.

The agriculture industry is a vital source of employment across the country. Agriculture also plays a vibrant role in the Indian economy. Agricultural managers form an essential part of the Indian-agro community, he or she manager promotes the resourceful production of quality food in the agricultural-food industry. Agricultural managers help farmers with improved plans for farming. An agricultural manager is instrumental in planning, organizing, and implementing farm management practices.

1. **Planning:** Planning for a farm comprises improving efficiency and taking decisions for the farm for its economic benefit. Things should be planned by the agricultural management for better forthcoming growth. An agricultural manager should plan and he/she should have the capability to perceive things from a future perspective.
2. **Making policy decisions:** The role of agricultural management is to create numerous policies concerning agriculture. The agricultural management makes policies essentially for the farmers, and villages dependent upon the area and the expected result from that particular area.
3. **Organising:** Agricultural management activities include daily planning, organizing, control and supervision, and administrative activities. This involves timely arrangement of required resources. An agricultural manager arranges for the agricultural equipment, crops production, agricultural products, and purchase and sales of livestock. He/she runs the farm efficiently and makes sure that the business does not face a financial downturn and remains profitable.
4. **Record keeping:** An agricultural manager is responsible for managing paperwork and keeping administrative records safe. Keeping the records in a proper way aids in analysing the performance of the farm. It also helps in knowledge management where previous records can be analysed to find a solution to an existing problem.
5. **Recruiting:** Agricultural management involves recruiting, training/instructing, and supervising farm workers. A farm manager is responsible for recruiting, training, and instructing farmworkers to achieve the desired outcome.
6. **Controlling and Supervision:** Agricultural management involves control over outcome by closely monitoring and supervising all the tasks to make sure that work is progressing satisfactorily. To ensure that the work progresses satisfactorily without any hitch, the farm manager, the farmers and other people working on the farm needs to coordinate and keep checking the work progress.
7. **Compliance:** Agricultural management is accountable for guaranteeing compliance with government regulations and health and safety standards. Compliance will ensure the smooth functioning of the farm as all rules and regulations are followed with utmost care.

8. **Financial planning:** Agricultural management involves ensuring that the farm is profitable and meets projected financial targets. Financial planning is an important aspect of agricultural management.
9. **Managing the farm:** Agricultural management involves organizing the maintenance/repair of farm property, equipment, and machinery. Differences in workloads can happen due to seasonal demands and managers will be expected to deal with emergencies.

Agricultural management practices are detailed procedures that are used in agriculture to create safe and wholesome food for consumers. A farmer performs major steps from crop selection to harvesting which includes Crop selection, Land preparation, Seed selection, Seed sowing, Irrigation, Crop growth, Fertilizing, and Harvesting. Good agronomical practices are the application of all these eight steps combined into farm management systems to improve soil quality, enhance water use, manage crop residues, and improve the environment through better fertilizer management.

1.4. Overview of Key International Standards such as ISO 26000 and SA8000, and their requirements in the context of Indian agriculture

ISO 26000 provides guidance to businesses and organizations which are committed to functioning in a socially responsible way. It is meant for organizations that understand the fact that respect for society and the environment is a critical success factor in growth. The application of ISO 26000 is progressively viewed as a way of evaluating an organization's commitment to sustainability. ISO 26000 is an international reference document for social responsibility, it provides best-practice guidance which helps organizations operate in a transparent and ethical manner contributing to sustainable development.

The food and agricultural industry have a huge role to play in the sustainable development of our world. The ISO/TS 26030 (Social responsibility and sustainable development) is a technical specification of ISO 26000 for the agri-food industry that offers sector-specific guidance on implementing ISO 26000 food and agricultural industry. It provides guiding principles on how an organization in the food production chain can contribute to sustainable development while considering all local laws, regulations and stakeholder expectations. These standards aid food companies, farms, cooperatives, processors and retailers, irrespective of their size or location, to develop a list of actions that will lead them to be more socially accountable.

ISO 26000:2010 in the food chain focuses on the seven core themes, namely organizational governance, human rights, environment, labour practices, consumer issues, fair operating practices, and community involvement and development. The foremost objective of these standards is to help organizations in the food chain industry, irrespective of their size or location. These standards draw a list of recommendations to move towards socially responsible behaviour. In India, the first CSR Foundation to receive the ISO 26000:2010 assurance is JSW Foundation. The social development wing of the JSW Group, JSW Foundation integrates social responsibility governance in accordance with ISO 26000 standards.

The SA8000 Standard is a global social certification program. The Standard and Certification System SA8000 provides a framework for organizations to conduct business in fair and decent manner. Compliance to SA8000 validate adherence to the highest social standards. This certification applies to organizations of all types irrespective of their industry or country. It is created by 'Social Accountability International' in 1997 as the first credible social certification framework. SA8000 applies management-systems approach to social performance and stresses on constant improvement. Child Labor, Forced or Compulsory Labor, Freedom of Association & Right to Collective Bargaining, Health and Safety, Disciplinary Practices, Discrimination, Working Hours, Remuneration and Management System are some elements of SA8000.

SA8000 is a certifiable standard on the other hand ISO 26000 provides guidance to companies. ISO 26000 is descriptive guidance and SA8000 includes management systems. Licious Is India's First Animal Protein Brand to receive SA8000 certification.

1.5. Best Practices and Case Studies of Successful Implementation of Social Policies and Management Systems in Indian Agriculture

1.5.1. Crop Cluster Development Programme in Haryana

Haryana is mainly an agrarian economy and agriculture is as an important source of employment and income for its rural population. The Haryana government launched the Crop Cluster Development Programme (CCDP) to increase agricultural productivity, promote sustainable farming practices, and improve rural livelihoods. Cluster Development Programme (CCDP) is a progressive initiative and it aims at forming crop clusters for adopting progressive farming practices.

Objectives of CCDP

1. Enable the formation of crop clusters of farmers involved in the farming of similar crops.

2. Encourage the acceptance of advanced agricultural methods, technologies, and best practices.
3. Improve market connections and access to Agri-infrastructure for enhanced value chain integration.
4. Strengthen livelihoods and income generation opportunities in rural areas through agricultural diversification.

Implementation Strategy

Identification and formation of crop clusters: The programme categorizes probable crop clusters based on factors like geographical proximity, crop specialization, and interest of the farmer. Farmers engaged in the cultivation of similar crops are encouraged to form clusters or cohesive groups.

Capacity building and training: The farmers in the crop clusters obtain training and capacity building programs on numerous facets of advanced farming practices, including crop selection, seed treatment, irrigation techniques, integrated pest management, soil health management, and post-harvest management.

Technology adoption: The programme encourages the implementation of modern technologies such as precision agriculture, mechanization, drip irrigation, greenhouse farming, and organic farming practices. To encourage the adoption of these technologies, it provides technical support and grants to farmers.

Market connections and infrastructure development: The CCDP enables improved market ease in forming linkages for farmers by establishing agri-marketing infrastructure, farmer-producer organizations, and cold storage services. It also inspires the formation of value-added enterprises and aids farmers in accessing government subsidies and schemes.

Monitoring and evaluation: For effective implementation, the programme comprises of a vigorous monitoring and evaluation framework. This framework evaluates the progress, identify bottlenecks, and make necessary changes in processes in order to bring effective outcomes.

Outcomes and Impact

Increased agricultural productivity: The acceptance of advanced farming practices has resulted in increased yields of the crops and enhanced productivity for farmers in the crop clusters. Improved access to quality seeds, efficient irrigation, and mechanization have contributed to positive outcomes.

Sustainable farming practices: The programme has promoted the adoption of sustainable farming practices, including organic farming, integrated pest management, and efficient water management techniques. These sustainable farm practices have led to reduced chemical usage, minimized environmental impact, and improved soil health.

Income generation and improvement of livelihood: The program has helped in diversifying rural livelihoods and generating additional income streams for farmers. Value addition through processing, branding, and marketing of agricultural produce has created employment opportunities and improved the economic conditions of rural communities.

Market integration and competitiveness: By establishing better market linkages, the CCDP has enabled farmers to access larger markets, obtain fair prices for their produce, and reduce dependency on intermediaries. The establishment of agri-infrastructure facilities has improved the value chain integration and enhanced the competitiveness of farmers.

Replicability and scalability: The success of the CCDP in Haryana has attracted attention from other states in India. The program has the probability to be replicated and scaled up across the country to benefit farmers on a larger scale.

The Crop Cluster Development Programme in Haryana has promoted development of crop clusters and inspired the adoption of advanced farming techniques. This program has created a favourable environment for agricultural development and innovation. It is an advanced initiative combining technological advancements, sustainable farming practices, and market integration to boost agricultural productivity and improve rural livelihoods.

1.5.2. Rythu Bharosa Kendralu - Farmer Assurance Centre

Agriculture is one of the dominant sectors in Andhra Pradesh and farmers provide food security to the whole country. In spite of technological progress in the agricultural sector, the farmers have relied on the advice of fellow farmers and input dealers. The farmers lack the residual knowledge of agronomical practices and the critical stages of various crops. They often pay an overprice for seeds, fertilizers, and other inputs, and these may not be recommended for the crop. As a result of using unsuitable resources, the farmers get deprived quality yields and are deprived of compensable prices, they are conned by middlemen and involuntary get into debt traps. To address the difficulties faced by the farmers, the Chief Minister of Andhra Pradesh established Rythu Bharosa Kendralu (RBK) in 2019. The literal meaning of Rythu Bharosa Kendralu which in State Language Telugu is 'Farmer assurance center'.

RBKs chiefly function through 4 verticals

1. Supply of pre-tested quality inputs: RBKs ensure reduced incidence of forged seeds and prevent the sale of inputs at excessive prices at private outlets. Inputs are made accessible ahead of the sowing season.
2. Delivery of services, capacity building, and knowledge dissemination: RBKs work towards farm mechanization and free crop insurance registration. They provide seed-to-seed training programs, training, and field diagnostic visits by scientists. They use posters, and pamphlets to spread information, and libraries with books and video materials on crop management are made available for information dissemination.
3. Farmers advisories and call center: Agri advisory board was established to assist in adopting best practices, providing support packages and addressing the domestic demand-supply gap. Integrated call centres are set up to facilitate instant remedies and instantaneous redressal of farmer grievances.
4. Procurement operations: The government has declared all RBKs as procurement centers to provide a platform for farmers to ensure MSP. It also reduces the transportation costs of the farmers.

Impact of RBKs

The scheme has aided to achieve higher yields and improved the quality of crops. It has ensured the supply of superior-quality agricultural inputs like seeds, fertilizers, and pesticides. The right advisories were provided at the right time to the farmers. As RBKS are declared to be procurement centers, farmers are now able to sell their supply in the village itself. Famers can now access government schemes through the RBKs and this has ensured transparency and speed. The RBK is an innovative and comprehensive approach. It is a Game Changer for farmers as it delivers all the services to the farmers from a single platform. It has strengthened the Agriculture Extension System in collaboration with all allied departments like Agriculture, Horticulture, Sericulture, Veterinary, Fisheries, Irrigation, etc.

1.6. The Way Forward for Policy Makers

Addressing the challenges of the agriculture sector is significant for India's overall development and growth. It is vital for the enhancement and welfare of the poor people living in rural India. Raising productivity per unit of land can fast-track agricultural growth as almost all cultivable land is being farmed. Because of the limitation of water resources, the use of water for irrigation/

must become more resourceful and efficient. Some other measures to increase productivity may include increasing yields, diversification to higher-value crops, and creating value chains to reduce marketing costs, etc.

Reducing rural poverty using a socially inclusive strategy is a key enabler for the growth of the country. Increasing both agriculture as well as non-agricultural income and employment must be a topmost priority. Rural development essentially should also benefit the poor, landless, women, scheduled castes, and tribes. There are strong regional disparities in India. the majority of India's poor are in rain-fed areas or in the Eastern Indo-Gangetic plains. Reaching such groups has not been easy. There is some progress with the rural population classified as poor fell from nearly 40% in the early 1990s to below 30% by the mid-2000s (about a 1% fall per year) – there is a stronger need for a faster reduction. Hence, poverty alleviation is a vital pillar of rural development.

Policymakers will thus need to initiate and/or conclude policy actions and public programs to shift the sector away from the existing policy and institutional regime and build a solid foundation for a much more productive, internationally competitive, and diversified agricultural sector. The policies areas should focus on;

1. Promoting new technologies and reforming agricultural research and extension: Brining major reform and strengthening of India's agricultural research and extension systems is a top priority for agricultural growth. Research and development in the areas of agricultural technologies, innovative approaches to farming, and the use of modern methods are necessary to increase productivity in the agricultural sector.
2. Improving Water Resources and Irrigation Management: Agriculture is India's largest user of water. As the demand for water multiplies for domestic and industrial use, less water is likely to be available for irrigation. The productivity of irrigation needs to be enhanced. Some ways of doing this can be improving on-farm management of water, and the use of more efficient delivery mechanisms such as drip irrigation. There is also a need to manage the use of groundwater by providing some incentives to pump less water can be thought of. Modernizing irrigation and drainage systems is key to the sustainable use of water.
3. Facilitating diversification to higher-value commodities: Inspiring farmers to diversify to higher-value commodities will be a significant factor for higher agricultural growth, mainly in rain-fed areas where poverty is high. The farmers and entrepreneurs should lead the

diversification initiatives and the liberalization of marketing, transport, export, and processing should be done by the Government. Also, great potential exists for increasing agro-processing and building competitive value chains from farmers to markets.

4. Developing markets, agricultural credit, and public expenditures: The agricultural sector in India has high-cost marketing and transport options for agricultural commodities. The private sector investment in marketing, value chains, and agro-processing is growing, but at a slower speed. Creating access to rural finance for farmers is another area to focus on. Moreover, have progressively come to dominate Government expenditures in the agricultural sector is dominated by subsidies on power, fertilizers, and irrigation. The government now needs to prioritize expenditure on agricultural research.

5. Poverty alleviation and community growth: Agricultural growth is vital for the overall development of the country. For the 'below poverty line' people living in rural India, additional measures are required to make the country's growth inclusive. Rural livelihood programs empower the communities to become self-reliant and have been found to be effective for scaling up. These programs promote the formation of self-help groups, increase community savings, and promote local initiatives to increase incomes and employment. By amalgamating to become larger entities, these institutions gain the strength to negotiate better prices and get market access for their farm produce and also gain the political and negotiating power over local governments to provide them with better technical and social services. These self-help groups are very effective at reaching women and poor marginalized families.

6. Sustaining the environment and future agricultural productivity: In some parts of India, the over-pumping of water for agricultural use is leading to falling groundwater levels. On the other hand, water logging in some areas is leading to the build-up of salts in the soils. In rain-fed areas on the other hand, where the majority of the rural population live, agricultural practices need to adapt techniques to reduce soil erosion and increase the absorption of rainfall. Damaged and degrading forest land require mitigation measures. Watershed management programs are a proven solution to these problems. In these programs, communities engage in land planning and adopt agricultural practices that protect soils, increase water absorption, and raise productivity through higher yields and crop diversification. These programs must be scaled up such initiatives to cover larger areas of the country. Weather change like droughts, floods,

and inconsistent rains must also be considered. They have extreme impact in rain-fed areas. The watershed program can be the most suited agricultural program for promoting new varieties of crops and improved farm practices. Other programs like the livelihoods program and the improvement of non-farm employment are also important in the growth of the agricultural sector.

A significant and sustained increase in farmers' income and the transformation of agricultural sector require a paradigm shift in the entire approach to agriculture. Changes in olden methods and policies are a must for creating an enabling environment for a modern and vibrant agriculture. Advancement in science led technology, an enhanced role of private sector in both pre and postharvest phases, liberalised market, and emphasis on efficiency will equip agriculture to address the challenges of 21st century and contribute towards the goal of a new India. Well-coordinated social policies between the centre and the states is required to ensure that agriculture industry moves to the next stage of development along with the other sectors.

Chapter Section Summaries

1.1. Overview of the Indian Agriculture sector and challenges faced: This section introduces readers to the Indian agriculture sector and provides an overview of its current state. The section identifies key challenges faced by the sector, such as farmer indebtedness, unequal distribution of resources, and limited access to technology and finance. The section also recognizes the significance of agriculture in India's economy and highlights the need for sustainable agriculture practices to ensure the sector's long-term growth.

1.2. Overview of the current state of social policies in Indian agriculture: In this section, readers gain an understanding of social policies in Indian agriculture. The section provides a brief history of agricultural policies in India and introduces current government schemes such as the Pradhan Mantri Fasal Bima Yojana, the National Agriculture Market, and the Soil Health Card Scheme. The section also examines the challenges faced in implementing social policies, including fragmented governance and inconsistencies in policy implementation across states.

1.3. Overview of Management Systems & Processes in the Indian Agricultural Context: This section discusses management systems and processes in Indian agriculture. The section identifies management systems such as Total Quality Management, ISO 9001, and Hazard Analysis and Critical Control Points, and explains how they can be utilized in agriculture to improve farming practices and enhance productivity. The section also outlines management processes such as procurement, logistics, and supply chain management, which can reduce wastage and increase yields in agriculture.

1.4. Overview of key international standards such as ISO 26000 and SA8000 and their requirements in the context of Indian agriculture: This section focuses on the significance of international standards like ISO 26000 and SA 8000 in the Indian agricultural context. The section explains the requirements of these standards concerning social responsibility and ethical business practices. The section also identifies examples of companies in India that have implemented these standards, demonstrating how they can help boost competitiveness and sustainability in agriculture.

1.5. Best practices and case studies of successful implementation of social policies and management systems in Indian agriculture: This section examines successful case studies of social policies and management systems in Indian agriculture. The section highlights the best practices utilized by these case studies, such as the adoption of precision agriculture technologies or the application of integrated pest management approaches. The section also showcases how these practices have resulted in positive social and environmental impacts, such as increased yields, income, and reduced pesticide usage.

1.6. The way forward for the policymakers: In this section, the author draws conclusions and recommends the way forward for policymakers in Indian agriculture. The section highlights the need for the development of policies that recognize the importance of sustainability, equitable distribution of resources, and technological advancement. The section also emphasizes the importance of efficient governance, proper implementation of policies, and the promotion of innovative business models to ensure the growth of the agriculture sector.

Discussion Questions

1. What are the current challenges faced by Indian agriculture today?
2. How would you explain the current state of social policies in Indian agriculture?
3. Why do you think implementing management systems is important for Indian agriculture?
4. How do ISO 26000 and SA8000 standards relate to Indian agriculture and why are they significant?

5. Based on best practices and case studies, how effective are social policies and management systems in improving Indian agriculture?
6. If you were a policymaker, what steps would you take to address the challenges faced by Indian agriculture and improve its future prospects?

Application Case Studies

Case Study 1: Crop Cluster Development Programme (CCDP)

Despite the government's efforts in implementing the Crop Cluster Development Programme (CCDP) in Haryana, the agricultural sector is affected by natural disasters such as droughts and floods, which reduce the farmer's productivity, affect crop yields, and affect their livelihoods. Farmers have had to adopt new methods, including the use of pesticides, herbicides, and fertilizers, which caused environmental degradation and increased the risk of crop failures. The CCDP program faced challenges in addressing these unpredictable setbacks, and the farmers who were expected to benefit from the programme were now faced with challenges that they could not handle alone. Many farmers struggled with paying back the loans taken, further worsening their poverty levels. The programme needed to address these challenges to ensure that it continued to be beneficial to farmers in Haryana.

Case Study Questions

1. How did the Crop Cluster Development Programme (CCDP) address the challenges of the use of pesticides, herbicides, and fertilizers in agricultural practices in Haryana?
2. To what extent has the CCDP contributed to reducing the negative impacts of natural disasters such as droughts and floods on farming practices in Haryana?
3. How did the CCDP programme respond to farmers' needs and address their challenges in loan repayment and overall poverty reduction?
4. What policy reforms were taken to help farmers adopt more sustainable farming practices in agriculture in Haryana, and how effective were these reforms?
5. How can the CCDP improve and ensure that natural disaster preparedness, environmental sustainability, and poverty alleviation are included in their long-term strategy to empower farmers in Haryana?

Case Study 2: Rythu Bharosa Kendralu (RBK)

The Indian agriculture industry faces challenges, particularly regarding small farmers, who are vulnerable to access to information and profitable markets. In Andhra Pradesh, farmers struggle for better yields and returns and are prone to fall into debts. In response to the issue, the Chief Minister established "Rythu Bharosa Kendralu (RBK)," which aimed to provide necessary agricultural support to these farmers. However, RBK faced many challenges, including corrupt practices in agricultural management, lack of proper communication channels, and inadequate staff with agro-experience. These challenged RBK's efficacy to deliver significant change to the Andhra Pradesh farmers.

Case Study Questions

1. What impact did the RBK initiative have on the lives of small farmers in Andhra Pradesh, and how was this measured or evaluated?
2. What were the aims of the RBK initiative, and how did it address the issues of farming practices in Andhra Pradesh?
3. What were the main challenges faced by RBK, and how did the organization address these challenges to ensure that their mission of empowering farmers was achieved?
4. What role can the Indian government play in supporting agricultural initiatives such as RBK, and how can they ensure that these programs are more effective in bringing about change?
5. What lessons can other regions in India learn from the RBK case study, and how can these be applied to other social policies and management systems within the agricultural context to achieve better outcomes for farmers?

References

Bajar, S and U Mushtaq (2019): "The processes of Transformation", in Inequality and the Demand for Non-Farm Jobs, NIAS Report: NIAS/SSc/IHD/U/RR/15/2019

Basole, Amit (2021a). Labour, Livelihoods, and Employment in the 2021–22 Union Budget. Economic & Political Weekly 56 (9):43-47.

Best practices in the social sector: A compendium 2023 retrieved from https://niti.gov.in/sites/default/files/2023-05/Best-Practices.pdf

Chand, R (2017): Doubling Farmers' Income: Rationale, Strategy, Prospects and Action Plan, NITI Policy Paper No.1/2017, National Institution for Transforming India, New Delhi.

Chand, R and J Jumrani (2013): "Food Security and Undernourishment in India: Assessment of Alternative Norms and the Income Effect", Indian Journal of Agriculture Economics, Vol. 68, No.1, pp. 39-53.

Chand (2022): "Agricultural Challenges and Policies for the 21st Century", NABARD Research and Policy Series No.6/2022

Chand, R, S K Srivastava and J Singh (2017): "Changes in Rural Economy of India, 1971 to 2012: Lessons for Job-led Growth", Economic & Political Weekly, December 30, Vol. 53, No. 52, pp. 64-71.

Department of Agriculture, Cooperation & Farmers' Welfare Annual Report https://agricoop.nic.in/Documents/annual-report-2020-21.pdf ISO/TS 26030:2019 "Social responsibility and sustainable development — Guidance on using ISO 26000:2010 in the food chain" retrieved from https://www.iso.org/standard/71624. html

Leipziger, D. (Ed.). (2009). SA8000: The First Decade: Implementation, Influence, and Impact (1st ed.). Routledge. https://doi.org/10.4324/9781351279840

Ministry of Agriculture & Farmers Welfare Press release "Doubling Farmer's Income" by PIB Delhi retrieved from https://pib.gov.in/PressReleasePage.aspx?PRID=1813688 National Statistical Office (2021a): Periodic Labour Force Survey: July 2019-June 2020, Annual Report, Ministry of Statistics and Programme Implementation, New Delhi.

Social Accountability International "SA8000 Standard" retrieved from https://sa-intl.org/programs/sa8000/ The World Bank Feature Story "India: Issues and Priorities for Agriculture" retrieved from https://www.worldbank.org/en/news/feature/2012/05/17/india-agriculture-issues- priorities

Virmani, A (2008): "Growth and Poverty: Policy Implications for Lagging States", Economic & Political Weekly, January 12, Vol. 43, No. 2, pp. 54-62

3

Intellectual Property Rights in Indian Agriculture Context

V. Selvalakshmi

Associate Professor & HOD, Vellammal College of Engineering and Technology

Learning Objectives

1.1. To understand about the background and history behind the intellectual property rights also to examine the interlinkage between IPR and agricultural technology.

1.2. To analyze the current statutory laws in existence in terms of IPR and in specific to agriculture-based technologies.

1.3. To evaluate the needs, benefits and challenges faced under Intellectual Property Rights.

1.4. To understand about the various classifications of IPR and also to study about its implications pertaining to agriculture.

1.5. To study about the various laws protecting the plants varieties and farmer's rights and live case studies in the field of IPR to gain knowledge, ways to protect the farmers against exploitation.

1.1. Introduction to Intellectual Property

Definition and background to Intellectual Property

Intellectual property rights (IPR) have been defined as ideas, inventions, and creative expressions based on which there is a public willingness to bestow the status of property. In order for the inventors or creators of that property to profit commercially from their creative endeavours or reputation, IPR grant them certain exclusive rights. One of the key factors influencing the development, transfer, and use of technology is IPR.

The main goal of intellectual property rights is to support inventions by encouraging their use and protection in order to support the growth of industries, which in turn supports technological innovation and the transfer and dissemination of technology. Intellectual Property Rights (IPR) play a crucial role in promoting innovation, protecting inventions, and ensuring fair competition in various industries. Maintaining intellectual property rights (IPR) is a crucial foundation for innovation and advancement in any industry. In order to promote ongoing investment in research and development and to bring cutting-edge technology to farmers, intellectual property protection is essential.

In the context of agriculture, IPR has gained significant importance in recent years. It encompasses the protection of plant varieties, genetic resources, and traditional knowledge associated with farming practices.

IPRin agriculture aims to incentivize research and development (R&D) in the seed industry, foster the development of new crop varieties, and safeguard the rights of farmers and breeders.

Brief History of Intellectual Property Rights

First international agreement, which was signed in Paris in 1883, aim was to protect industrial property and also to assist inventors in ensuring that their intellectual works would be safeguarded in other nations. Literary (such as novels, short stories, poems, plays, etc.), musical (such as songs), and artistic works (such as drawings, paintings, sculptures, and architectural works, etc.) were all protected at the Berne Convention in 1886 with the aim of granting creators the right to oversee and be compensated for their creative works on a global scale.

The first worldwide IP filing service was introduced during the Madrid Agreement (1891). After two years, Berne saw the creation of the United International Bureaux for the Protection of Intellectual Property (BIPRI). In 1970, BIPRI was renamed as the World Intellectual Property Organisation (WIPO), an intergovernmental organisation with headquarters in Geneva. In 1974, the WIPO was included into the United Nations (UN), where all UN members had the option to join but were not required to do so.

In 1978, the Patent Cooperation Treaty (PCT) went into effect. The WIPO Arbitration and Mediation Centre was founded in 1994 to resolve international business disputes between private parties. The WIPO Academy was established in 1998 to offer introductory and advanced IP training. WIPO approved the Development Agenda in 2007 to ensure that the organisation will take on the development challenges. IPR became an important part of the World Trade Organisation (WTO) as a result of the standard laws and norms of patents, trademarks, copyright, etc. that were designed to reap the benefits of innovations.

However, the British government first implemented the patent system in India in 1856.In India, IPRs are protected by well-established statutory, administrative, and judicial frameworks. In addition, there are numerous lawyers and law firms in India's various states that deal with intellectual property. It is important to note that by passing the appropriate laws and revising the existing ones, India has fulfilled with its commitments under the Agreement on Trade Related Intellectual Property Rights (TRIPS). The protection of computer databases and software under India's copyright laws has successfully reduced piracy through judicial action. The 14-year patent period under the Indian Patent and Designs Act (1911) was administered by the Controller of Patents.

The Patents (Amendment) Rules, 2016, have revised the Indian Patents Rules, 2003. The modification is effective as of May 16, 2016. The current revision is significant and may undoubtedly affect, among other things, the patent application and prosecution process.

However, in general, the Patent Act, 1970, as amended in 1999 and 2005, as well as the Copyright Act, 1957, as amended in 1982, 1984, 1992, 1994, and 1999, as well as the Trademark Act, 1999, as amended in 2000, govern functional and technical inventions, purely artistic works, and a symbol, logo, work, sound, colour, design, etc.

Intellectual Property Rights and Agriculture Technology

Machines have traditionally been seen as a type of invention or artistic production that is entitled to intellectual property rights protection. However, in developed nations, the assignment of IPRs to living beings is a comparatively new development. The first vegetable propagated plants in the US were granted patentability in 1930. Plant breeders' rights (PBRs), a new type of intellectual property for the protection of plant varieties, originated in the second part of the 20th century. According to Indian law, the Protection of Plant Varieties and Farmers' Rights Act of 2001 protects IPRs relating to crop and planting innovation.

This law aims to improve plant variety protection, protect farmers' and plant breeders' rights, and promote the creation of new plant varieties. However, the length of a registered variety's protection depends on the sorts of crops. For example, the protection period for trees and vines is 18 years, for other crops it is 15 years, and for current varieties it is 15 years from the date of notification. So, in a way, we may argue that the plant protection system is expected to increase economic gains in the industrialised world. Private breeders are drawn to this kind of system because it safeguards their intellectual property. As an illustration, patents or PBRs typically place limitations on farmers' ability to sell grown seed, sometimes to reuse it, which increases the market for the breeder's seed.

Even in developed countries, though, it is still relatively usual to reuse seeds, even though annual purchases are now the norm for many crops. Contrarily, farmers in developing nations have customarily replanted, traded, or sold seed from the crop from the previous year. Most farmers in underdeveloped nations repurpose, trade, or sell informal seed to their neighbours; annual seed purchases are generally uncommon. Breeders obviously struggle to recover their investments in better varieties through recurring sales, which puts farmers and breeders in opposing perspectives.

The relationship between agriculture and IPR should be examined from another aspect. Three general sections make up this as mentioned in Fig.No:1 is

a) Technologies that impact IPRs

b) Technologies that are driven by IPRs (requirements of the IPR regime)

c) Technologies that are enabled by IPRs (incentives for current efforts).

It is entangled with one another. Plant variety development is influenced by the IPR regime's nature, the amount invested, the type of technology, and the rate at which it is created and transferred. In other words, the IPR regime will be one of the primary motivators to engage in plant protection or other types of crop cultivation-related activities. Different IPR types will control different technologies.

Gaining deeper insights into IPRs and related factors is unavoidable in order to visualise the future scenario in agricultural R&D. The combination of the IPR regime and the process of technology development and transfer would have numerous consequences for farmers, researchers, and agribusiness organisations.

As depicted in the figure 1, the relationship between IPRs and agriculture, demonstrating the dynamics of interaction in three ways that aid R&D and policy framework. The Indian Patent Act of 1970, together with GATT participation in 1987, established an IPR system in the form of several amendments and bills (for example, the Patents (Amendment) Act of 1994, the Geographical Appellation Bill of 1999, the Patents Bill of 1999, the PVP and Farmers Rights Bill of 1999, and so on). Currently, WTO members recognise the Agreement on Trade-Related Aspects of Intellectual Property Rights (TRIPS Agreement), which provides a number of precise minimum levels of protection.Technology has an impact on the social, economic, and ecological systems.

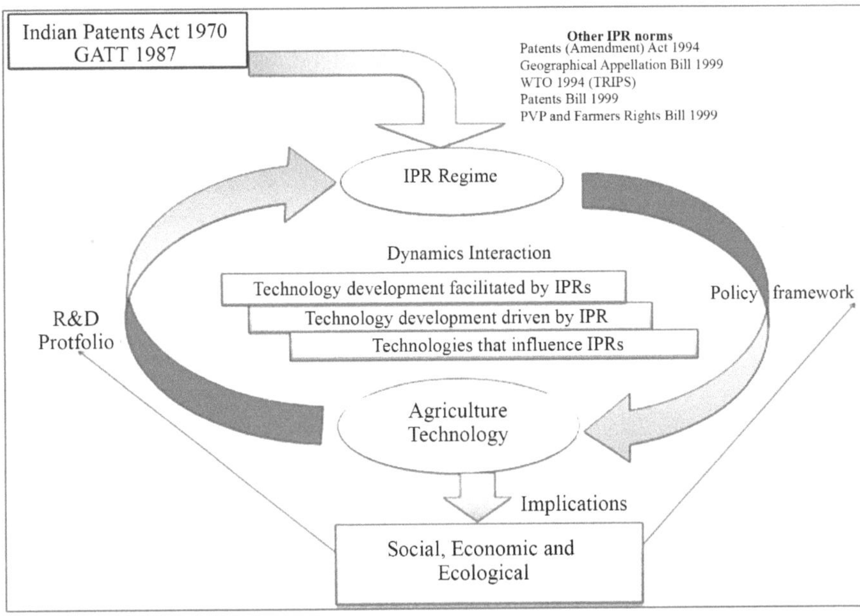

Fig. 1: IPR regime and technological processes in agriculture Source: Ravishankar and Archak, 2000

1.2. Current Statutory IP Laws in India vis-vis Agri-based Technologies in India

All members of the WTO, including India, are bound by the 1995 WTO-TRIPS agreement, which established minimum requirements and standards for the protection of intellectual property rights (IPR) in a number of categories, including patents, copyrights, trademarks, plant varieties, geographical indications, industrial designs, layout designs of integrated circuits, and trade secrets. India implemented a number of suitable and compliant methods and instruments as a result of this agreement. The Patents Act, 1970 (39 of 1970), The Patents (Amendment) Act, 1999 (17 of 1999), The Patents (Amendment) Act, 2002 (38 of 2002), and The Patents (Amendment) Act, 2005 (15 of 2005) are a few of the legal documents adopted by the Indian Parliament as part of the TRIPS compliance process. The Protection of Plant Varieties and Farmers Rights Act, 2001 (PPV FR Act) (53 of 2001) and the Geographical Indications of Goods (Registration & Protection) Act, 1999 (Office of Controller General of Patents Designs and Trade Marks, 2016). In addition to this, the Indian government passed law known as the Biological Diversity Act, 2002 (No. 18 of 2003). As part of the nation's dedication to the Convention on Biological Diversity (CBD), (NBA, 2008). There is no separate IPR Act that offers protection for trade secrets and other nonpublic information. This is covered by the Indian Contract Act of 1872 and common law, with the Ministry of Law and Justice serving as the nodal agency (Sudhir Kochhar, 2008).

The Annexure table given below lists the main Intellectual Property asset categories used in agriculture R&D together with the characteristics that make them eligible for protection under Indian law. Agri value chain stakeholders' potential rewards, broad institutional frameworks, and legal restrictions are also shown. The numerous laws and legal systems for protecting bioresources and traditional knowledge are also listed in light of the unique nature of their use in agriculture.

Annexure: Current Statutory IP Laws in India vis-vis Agri-based Technologies in India:

S.No	IP	Legislation	Administrationauthority	Qualifyingattributes	Potential stakeholder (s) to benefit
1.	Patent	Patents Act,1970 and amendments thereon	Controller General of Intellectual Property Rights	Novel, non-obvious, capable of industrial application and not fall withinthe provisions of Section 3 and 4 of the Patents Act, 1970; Also Statutory Compliance with Section 10 for those innovators based on biological resources	Inventors, traders, economy
2.	Design	Design Act,2000and amendments thereon	Controller General of Intellectual Property Rights	New or original; significantly distinguishable from knowndesigns or combination ofknown designs	Industry;SMEs
3.	Trademark	Trade MarksAct, 1999and amendments thereon	Controller General of Intellectual Property Rights	Capable ofdistinguishing features of goods andservices, capable ofgraphical representation, used or proposed to be used to identifygoods/services	Industry/SME: products andservice sector
4.	Geographical indication	Geographica l Indications (Registrationand Protection) Act, 1999	Controller General of Intellectual Property Rights	Specific geographical origin, possessing qualities, reputation orcharacteristics that are essentially attributable tothat place oforigin	Communities,tradi-tionalpractitioners, knowledgeholders
5.	Copyright	Copyright Act, 1957and amendmentsthereon	Controller General of Intellectual Property Rights	Original expressions ofideas, creations	Creators of allworks
6.	Integrated circuit design	Semiconductor Integrated Circuits Layout- Design Act,2000	Controller General of Intellectual Property Rights	Original; notcommercially Exploited anywhere in India or in a convention country; inherently distinctive; inherently capable of being distinguishable from any other registered layout-design	Electronic industry, traders, SMEs

7.	Plant varieties	Protection of Plant Varieties and Farmers" Rights Act, 2001	Chairperson, Protection of Plant Varieties and Farmers" Rights (PPV&FR) Authority/ Registry	New, distinct, stable and uniform	Plant breeders, farmers,industry
8.	Biodiversity	Biological Diversity Act, 2002	Chairperson, National Biodiversity Authority (NBA)	Biological resources, herbal remedies, associated traditional knowledge	Knowledge holders, farmers, communities, researchers, etc.
9.	Traditional knowledge	None	Secretary of the concerned Ministry (ies)	Traditional knowledge/ genetic resources	Knowledge holders in communities by sharingof accrued knowledge

*Kalpana Sastry,R,Intellectual Property Rights In Indian Agriculture, 105 FoCARS Foundation Course For Agricultural Research Service, ICAR-National Academy of Agricultural Research Management

Need for Intellectual Property

- The preservation of IPRs is crucial.
- Inventors are granted exclusive rights to exploit their intellectual property (IP) for their own financial and practical gain. The pricing of the innovation, the marketing and distribution methods, and a myriad of other ways to make money and get a good return on investment are all completely up to the creator.
- The innovator can prevent others from profiting financially by filing for IP.
- The right to sue the party trying to profit from the innovation can be exercised by the inventor if a rival violates the IP rules that protect innovators.
- If proven guilty, the court may grant the inventor a monetary reward, which would be paid by the person or organisation that was attempting to profit financially from the invention that was claimed to belong to the IP owner.
- IP protections are crucial because they give innovators the motivation to develop new ideas and profit from their creations.

1.3. Benefits of IPR

Intellectual property rights not only protect the ideas or concepts of business but also protect the genuine business assets that are vital to the products and services. Securing intellectual property rights has a number of benefits, including:

- Increases market value - IPRs can help you grow your firm through licencing, sales, and even commercialisation of the goods and services they have protected. In the end, this will increase market share and contribute to higher earnings. Intellectual property rights that have been registered and safeguarded can help increase a company's value in the event of a sale, merger, or acquisition.
- Transform thoughts and ideas into assets that generate income - Thoughts and ideas don't have much value on their own, but you can turn them into economically viable goods and services by registering them as intellectual property.
- It is possible to generate a consistent flow of royalties and additional money by copyrighting or licencing the patents.

- Promote your goods and services – Obtaining intellectual property rights can improve the reputation of your company. Your ability to distinguish your goods and services from competitors can be aided by intellectual property rights such as trademark registration.
- Open up or raise Capital - An individual can monetize for debt financing by sales, licencing, or by utilising IPRs as collateral. When requesting government financing such as grants, subsidies, and loans, intellectual property rights can be a benefit.
- Increases export potential - A company with registered IPRs will be able to advertise its goods and services to foreign markets by using brands and designs. A company can also take advantage of franchise deals with foreign firms or export trademarked goods.

1.4. Challenges

In the area of intellectual property rights (IPR), India has a number of difficulties that have an impact on its markets, innovation, and economic expansion.

Among the biggest obstacles are

- **Lack of Knowledge:** In India, many people and companies, particularly those in rural regions, are unaware of the value of IPR and how to safeguard their intellectual property. Low levels of copyright, trademark, and patent registration are the result of this.
- **Weak Enforcement:** India's legal system for protecting IPRs is regarded as being ineffective, and its enforcement procedures are insufficient. In especially in the software, pharmaceutical, and entertainment industries, this has resulted in significant amounts of piracy, counterfeiting, and IPR infringement.
- **Slow and Complex Legal Procedures:** The legal processes in India for acquiring and enforcing IPR are sometimes drawn out and complicated, which causes delays and expensive expenses for people and companies looking to protect their intellectual property.
- **Patent quality:** India has come under fire for issuing subpar patents that aren't original or inventive. This has raised worries about patent thickets, where numerous patents protect a single invention and cause legal wrangling and innovation-stifling litigation.
- **Access to Medicines:** India has been a significant producer of generic medications, and there are fears that increased IPR protection may restrict millions of people's access to reasonably priced medications.

1.5. Classification of IPR and its implications in Agriculture

On the basis of type of invention and creation of human mind and their application, the IPRs are classified as follows:

1. Patents - An invention is defined as a product or a process that offers, generally, a new way of doing something or offers a new technical solution to a problem. Any invention must meet certain requirements, such as Usefulness, Novelty, and Non obviousness, in order to be eligible for a patent, which provides protection for the invention to the owner of the patent for a finite period, i.e. 20 years.

Three fundamental statutory types of patents are utility, design and plant patent.

- A utility patent is a common type, which is given for any novel and useful process, machine, manufacturing, or composition of matter, or for any novel and useful enhancement thereof. It must be USEFUL, to put it simply.
- A design patent safeguards a brand-new, unique, and ornamental manufacturing design.
- A plant patent is awarded to protect a novel and unique asexually reproduced plant variety.

2. Trademarks - A trademark is a distinguishing symbol used to identify a certain person or business as the producer or provider of a particular good or service.1 It could consist of one word or a string of letters, numbers, or all three.

Concerns about branding and trademark protection towards the global agricultural sector includes following:

- The impact of agricultural product branding on markets
- The expansion of global marketing
- Brand image and quality assurance
- Using locally derived terminology/indigenous terms in a trademark, such as Basmati or Jasmine
- Branding-related expenses/costs

3. Copyright and related rights - Copyright is a legal word that refers to the permissions granted to authors and artists for their creative works. In exchange for cash, authors frequently sell the rights to their creations to people or businesses who can best promote them. These payments—which are subsequently known as royalties—are frequently made in accordance with the actual use of the work.

Further Copyright safeguards the expression of an idea, not the idea itself, in contrast to a patent, which defends an idea and how it is put into practice. Such expression must be recorded in a form that can be read, such as handwriting, typeset text, magnetic tape, or another type of storage medium. The expression in computer programmes, video or motion pictures, sound recordings, photography, and sculpture is protected by copyright. In contrast to patents, copyrights are created automatically once an idea is fixed in a material form of expression.

There are numerous challenges arising from this area of IPR protection due to the application of copyrights to protect other material essential to agriculture, such as genome databases, meteorological data, GIS imagery, etc.

The following significant topics will require more discussion:

1. Database copyright and its effects on genomics.
2. What is your view on "fair use" and how does it affect research?
3. International conventions on copyright enforcement
4. How much alteration is necessary to show that the information is new?
5. The internet's effects on copyright and fair usage

4. Geographic indications of source

A Geographical Indication (GI) is a label placed on products with a particular geographical origin and qualities, a reputation, or other traits that may be directly linked to that location of origin. HandiCrafts may be registered as Geographical Indications (GI) under the Geographical Indications of Goods (Registration and Protection) Act of 1999 (the Act). 1152 GI Authorised users were registered under the Act as of July 31, 2016, for handmade products.

The agricultural sector uses geographic indications (GI) more frequently than the industrial sector. These are logos associated with commodities from a country, region, or locale where the characteristics of the good are mostly determined by where it was produced. Numerous GIs have a connection to or come from agricultural items. Famous examples of these are found in that district's "Darjeeling for tea," "Devgad or Ratnagiri for mangoes," and "Tasgaon for grapes." In order to avoid being protected from protection for an extended length of time, such as that provided by plant patents or the rights of plant breeders, plant species developed using traditional knowledge and related to a particular location may also be considered to be GI.

5. Industrial designs - Industrial designs are works of art that give products an attractive or formal appearance, and "design rights" are privileges granted to the owner of a legally registered design over new or original designs. The New Designs Act, 2000, which intends to implement a more precise classification of design to comply with the worldwide system and to address the expansion of design-related activities in numerous domains, contains the current legislation on industrial designs in India.

6. Trade Secrets - In general, a trade secret is any exclusive commercial information that gives a company a competitive advantage. Manufacturing, industrial, and business secrets are all included in trade secrets. It is considered an unfair practise and a breach of the trade secret when such knowledge is used without authorization by parties other than the holder. Trade secrets are protected without registration, in contrast to patents.

For example, in agriculture, hybrid plant kinds may benefit from trade secret protection. Therefore, as long as they are kept secret, the use of hybrids is appropriate even in countries that do not recognise the rights of crop breeders. Laws prohibiting unfair competition, limiting business practises, or contract law may protect commercial secrets from unauthorised use by third parties. Agricultural, industrial, and utility products are all eligible to use trade marks. Trademarks may be used, for instance, to sell seeds or provide spraying services. A trademark's primary purpose is to distinguish products and services from one business and to keep customers from becoming dissatisfied.

7. Protection of plant varieties & farmer's rights : The purpose of this act is to reward traditional, rural, and tribal communities for their contributions to the country's agro-biodiversity and to recognise farmers' roles as cultivators and conservers of that diversity. It also aims to encourage investment in R&D for the creation of new plant varieties to support the expansion of the seed industry, which will ensure that farmers have access to high-quality seeds and planting materials. This will be discussed in detail in the later part of this chapter.

8. Protection of Biological Diversity - Traditional knowledge is covered under the Biological Diversity Act directly in the preamble. It also addresses traditional knowledge-related issues under the umbrella of linked knowledge in accordance with various requirements of the Biological Diversity Act, 2002. The people who are claiming benefits are those that protect biological resources, produce knowledge about their uses, and hold this knowledge and information.

Governing Regulations in IPR

Intellectual properties rights in India is governed under the following Acts:

- Trade Marks Act, 1999
- The Patents Act, 1970 (amended in 2005)
- The Copyright Act, 1957
- The Designs Act, 2000
- The Geographical Indication of Goods (Registration and Protection) Act, 1999
- The Protection of Plant Varieties and Farmers Rights Act, 2001
- The Information Technology Act, 2000
- The IPR security for fresh plant species in India in the form of the Protection of Plant Varieties and Farmers' Rights Act (PPVFR) of 2001

1.6. Overview of Indian Agriculture

Indian agriculture is the backbone of the country's economy, employing a significant portion of the population and contributing to food security. The agricultural sector in India is diverse, with a wide range of crops and farming practices. Small-holder farmers play a crucial role in this sector, relying on traditional knowledge, seed saving, and exchange practices to sustain their livelihoods. However, the sector faces numerous challenges, including low productivity, climate change, and the need for technological advancements. About 55% of India's population relies on agriculture as their main source of income, making India one of the leading players in the global agricultural industry. India is the world's largest producer of milk, pulses, and spices. It also has the largest herd of cattle (buffaloes), the largest area planted to wheat, rice, and cotton. It is the second-largest producer of wheat, rice, cotton, sugar, farmed fish, fruit, vegetables, tea, and farmed vegetables. About half of the population of India is employed in agriculture, which has the second-largest agricultural land area in the world. As a result, farmers play a crucial role in the industry that produces food for us.

One of the largest sectors in India, food processing contributes 32% of the nation's total food market and is rated fifth in terms of production, consumption, export, and expected growth. India's agricultural machinery market, which was worth Rs. 908 billion in 2018, has significant growth and development potential. The benefits of farm productivity mechanisation and credit facilities that are available to farmers have been made possible by the need for agricultural gear. Intellectual property is becoming more and more important in the agricultural industry today.

1.7. Protection of Plant Varieties and Farmers' Rights Act (PPV & FR), 2001

One of the significant act under IPR in agriculture is Protection of Plant Varieties and Farmers' Rights Act, 2001. It has been considered necessary to recognise and protect the rights of the farmers in respect of their contributions made at any time in conserving, improving, and making plant genetic resources available for the development of new plant varieties in order to facilitate the establishment of an effective system for the protection of plant varieties, the rights of farmers, and plant breeders. "The Protection of Plant Varieties and Farmers' Rights (PPV&FR) Act, 2001" was passed by the Indian government using a sui generis method.

The laws of India are not only compatible with the International Union for the Protection of New Varieties of Plants (UPOV), 1978, but they also contain enough protections for farmers' and public sector breeding institutes' interests.

The law acknowledges the roles played by farmers and commercial plant breeders in plant breeding activities and calls for TRIPs to be implemented in a way that advances the unique socioeconomic interests of all parties involved, including the public and private sectors, academic institutions, and farmers with limited resources.

1.8. History of the Protection of Plant Varieties and Farmers' Rights Act

After India joined the World Trade Organisation in 1995 and agreed to implement the Agreement on Trade-Related Aspects of Intellectual Property Rights (TRIPS), there were heated discussions about how intellectual property rights should be incorporated into Indian agriculture, which resulted in the introduction of the Protection of Plant Varieties and Farmers' Rights Act in 2001.

India had to decide whether to approve a law that protected cultivating networks' interests or to accept the privileges granted to plant breeders by the International Union for Protection of New Plant Varieties (UPOV Convention). The later option was essentially rejected because the farmers were not allowed to reuse ranch leftover seeds or trade them with their neighbours under the current adaption of UPOV, which was adopted in 1991.

India added a section on farmers' rights to the PPVFRA that has three pillars:

1) Farmers are recognised as plant breeders and can register their varieties;
2) Farmers who conserve landraces and wild relatives of economic plants and improve them through selection and preservation are recognised for their work and rewarded; and
3) Farmers' traditional practises of saving seeds from one harvest, are safeguarded and they can share with their farm neighbours.

PepsiCo case

- Two hybrid potato varieties, FL 1867 and FL 2027, were submitted for registration by pepsi co in February 2011.
- In February 2016 the PPVFRA approved the registration of these varieties for a 15-year period. The hybrid varieties were later marketed by PepsiCo under the name FC-5, and the company alleged that Gujarati farmers were using it unlawfully.
- But as long as certain requirements are followed, the PPVFRA permits farmers to reuse registered varieties and share them with their neighbours, therefore doing so is not illegal per se.
- The company asserted in court that farmers may bring potatoes of the specified variety on the buyback system since FC-5 had a licence. Without breaking the law, the FC-5 type could have been made available and disseminated everywhere.
- The FC-5 was additionally listed as a "Extant Variety," which is another name for a "Variety of Common Knowledge." It suggests that the aforementioned potato variety was "common knowledge" in the nation before it was registered and that it was already available in the country before it was.

1.9. Objectives of the Protection of Plant Varieties and Farmers' Rights Act, 2001

1. To create a framework that effectively protects plant varieties, farmer and plant breeder rights, and to promote the creation of new plant varieties.
2. To acknowledge and defend farmers' rights with regard to their ongoing efforts to preserve, enhance, and make available plant genetic resources for the creation of new plant kinds.

3. Protecting the rights of plant breeders will help the nation's agricultural development along with encouraging public and private sector investment in research and development to create new plant types.
4. Encourage the development of the nation's seed business, which will guarantee that farmers have access to high-quality seeds and planting supplies.

Authority

The Protection of Plant Varieties and Farmers' Rights Authority will be established by the central government. A director and fifteen representatives from various relevant services and offices, the seed business, farmers groups, inborn networks, the State-level Ladies' Association, etc. will make up the group.

Rights under the PPV&FR ACT 2001

- **Breeders' Rights:** The protected variety may only be produced, sold, marketed, distributed, imported, or exported by breeders. Breeder may designate an agent or licensee and pursue a civil remedy in the event that their rights are violated.
- **Researchers' Rights:** Under the Act, researchers may conduct experiments or conduct research using any of the registered varieties. This includes using a variety as a starting point for the development of another variety, but repeated use requires prior consent from the registered breeder.
- **Farmers Rights:** Farmers' varieties can also be registered as existing varieties. A farmer who has created a new variety is entitled to registration and protection in the same way as a breeder of a variety.

Farmers are eligible for recognition and rewards for the conservation of Plant Genetic Resources of Land Races and Wild Relatives of Economically Important Plants (PPV&FR Act, 2001); Farmers are allowed to save, use, sow, re-sow, exchange, share, or sell their farm products, including seed of a variety protected under the Act; Farmers shall not be allowed to sell branded seed of a variety protected under the Act.

1.10. Registration of Varieties

A variety may be registered under the Act if it essentially satisfies the Distinctiveness, Uniformity, and Stability (DUS) requirements. For the aim of registering varieties, the Central Government publishes notifications in official gazettes naming the genera and species. The Central Government has announced 157 crop species for registration thus far.

For each crop species, the PPV&FR Authority has created "Specific Guidelines" or "Guidelines for the Conduct of Species Specific Distinctiveness, Uniformity and Stability" tests.

Registration fees under PPV&FR ACT 2001

Plant variety registration applications must be accompanied by the registration fee specified by the authority. The following are the registration costs for the various varieties:

S.No	Types of Variety	Fees for Registration
1	Extant Variety notified under section 5 of the Seeds Act,1966	Rs 2000/-
2.	New Variety/Essentially Derived Variety (EDV)/ Extant Variety about which there is common knowledge (VCK)	• Individual Rs. 7000/- • Educational Rs.10000/- • Commercial Rs.50000/-
3.	Farmers Varieties	No Fee

Source: https://plantauthority.gov.in/

1.11. International Union for Protection of New Plant Varieties (UPOV Convention)

- An intergovernmental organisation called the International Union for the Protection of New Varieties of Plants (UPOV) is based in Geneva, Switzerland.
- The International Convention for the Protection of New Varieties of Plants established UPOV. In addition to being updated in 1972, 1978, and 1991, the Convention was approved in Paris in 1961.
- The goal of UPOV is to encourage the production of new plant varieties for societal benefit by providing and promoting an efficient system of plant variety protection.
- The UPOV Convention establishes the framework for members to support plant breeding by granting the breeder's right—an intellectual property right—to those who create new plant varieties.
- When a variety is covered by a breeder's right, commercial propagation of the variety requires the breeder's consent. The individual UPOV members grant the breeder's right.

- An innovative plant variety can only be safeguarded by its breeder. A variety's protection cannot be obtained by anyone other than the breeder.
- Under the UPOV system, there are no limitations on who can be regarded as a breeder: a breeder could be a person, a farmer, a researcher, a governmental institute, a private corporation, etc.
- India is not a member.

Further Roadmap

IPR in agriculture prohibits third parties from misappropriating the produce of farmers, breeders, and others and reaping unfair commercial advantages by providing legal protection and remedies in the event of a violation of the rights assigned to the actual owners.

Strengthening the Policy Area enables the protection of IPR in agriculture in better way. Recognising that the competitive commercialization of technology is the primary policy issue connected to IPR protection in agriculture and associated sectors, focus should be towards further liberalisation of agricultural markets and promotion of private investment in the sector and more efficient technology systems.

Recognising the importance of increasing farm enterprise support services, extension, training, research, and quality control, public interventions in agriculture should focus on market information, technology forecasting, and early warning systems. To improve the prospects and sustainability of competitive Indian agriculture, a centre for anticipating market trends and the state of national and worldwide markets should be built. Marketed innovations should be created, protected, and commercialised in order to maximise returns on investments.

To take advantage of the most IPR-related potential in competitive agriculture, India must continue to contribute to the creation of a fair playing field at the intergovernmental platform between developing and developed nations. In the upcoming World Intellectual Property Organization (WIPO) talks for IPR in connection to genetic resources, traditional knowledge, and traditions, NARS specialists may play an important role, which the government may capitalise on.

Chapter Section Summaries

Section1. To understand about the background and evolution of intellectual property rights. This chapter also provides the overview about relationship between IPRs and agriculture, demonstrating the dynamics of interaction in that aid R&D and policy framework.

Section 2. To analyze the current statutory laws in existence in terms of IPR and in specific to agriculture-based technologies. This also highlights various Institutional mechanisms, legislative provisions for specific Intellectual Property and related forms of knowledge and resources for agricultural technologies.

Section 3. To evaluate the needs, benefits and challenges faced under Intellectual Property Rights. In this section, needs for IPR, benefits and various challenges faced during the implementation of IPR is examined in detail.

Section 4. To understand about the various classifications of IPR and also to study about its implications pertaining to agriculture. This section also highlights various governing regulation with regards to IPR and a learning case is also been discussed.

Section 5. To study about the various laws protecting the plants varieties and farmer's rights , the rights and regulations and authority are discussed and live case studies in the field of IPR with reference to agriculture to gain knowledge to protect exploitation of IPR rights is also discussed. Further roadmap is also discussed.

Application Case Studies

Case Study 1

Syngenta v. Bunge (2004) was a notable legal case involving patent infringement and the unauthorized sale and export of genetically modified (GM) corn. The lawsuit was initiated by Syngenta, a multinational agribusiness company, against Bunge, a leading grain trading company.

In the early 2000s, Syngenta had developed and obtained a patent for a specific genetically modified corn variety known as "Event 176." This GM corn was engineered to possess certain desirable traits, such as resistance to pests and herbicides, which made it highly valuable in the agricultural industry. Syngenta had established licensing agreements with various companies, allowing them to legally produce and sell seeds of Event 176.However, Syngenta alleged that Bunge had engaged in the unauthorized sale and export of Event 176 corn without obtaining the necessary licensing agreement. According to Syngenta, Bunge had acquired the GM corn through indirect channels, such as purchasing it from farmers who had illegally saved and replanted Event 176 seeds without permission. Syngenta contended that Bunge had knowingly participated in this unauthorized distribution chain and profited from it.

The case revolved around the issue of patent infringement. Syngenta argued that Bunge's actions violated its exclusive rights as the patent holder, including the rights to produce, sell, and distribute Event 176 seeds. Syngenta sought damages for the alleged infringement, claiming that Bunge's actions had caused financial harm to the company.

The legal proceedings Involved a thorough examination of the evidence, including the documentation of seed purchases, sales records, and communications between Bunge and the farmers. Expert witnesses were called upon to provide testimony regarding the genetic characteristics of the corn and the potential impact of unauthorized distribution on Syngenta's business interests.

As the case unfolded, several questions came to the forefront:

1. Did Bunge knowingly engage in the unauthorized sale and export of Syngenta's genetically modified corn?
2. To what extent can Bunge be held liable for the actions of farmers who illegally saved and replanted Event 176 seeds?
3. What is the scope of Syngenta's patent rights, and did Bunge's actions fall within the bounds of patent infringement?
4. What damages, if any, should be awarded to Syngenta for the alleged infringement?

The case attracted considerable attention within the agricultural and legal communities due to its potential implications for intellectual property rights and the enforcement of patent protections in the field of biotechnology. It also underscored the importance of licensing agreements and proper oversight in the commercialization of genetically modified crops.

Ultimately, the outcome of Syngenta v. Bunge (2004) was not available within the provided information. However, this case serves as a significant example of the legal complexities surrounding patent infringement and the protection of intellectual property in the agricultural industry.

Case Study 2

Farmers' Seed Network v. Monsanto (2011) was a significant legal case that involved a dispute between Farmers' Seed Network, an organization advocating for farmers' rights, and Monsanto, a multinational agribusiness company. The lawsuit centered around the patent rights of genetically modified seeds and the ability of farmers to save, exchange, and replant those seeds without infringing on Monsanto's patents.Farmers' Seed Network argued that Monsanto's patent rights should not extend to the progeny or subsequent generations of genetically modified seeds when saved and replanted by farmers. The organization contended that once the patented seeds were purchased and planted by farmers, they should have the freedom to save and reuse the seeds without restrictions, as has been a common practice in traditional agriculture for centuries.

The case raised several important questions:

1. Can farmers save and replant genetically modified seeds without infringing on Monsanto's patent rights?
2. Should Monsanto's patent rights extend to the offspring of genetically modified seeds?
3. What are the implications of allowing farmers to save and replant patented seeds without licensing agreements?
4. How does this case impact the balance between protecting intellectual property rights and promoting agricultural practices that have traditionally involved seed saving and exchange?

Ultimately, the court ruled in favor of Monsanto, upholding the company's patent rights. The court held that the act of saving and replanting genetically modified seeds constituted patent infringement, regardless of whether it was the original patented seed or its progeny. The ruling maintained the existing legal framework that protected Monsanto's rights to its genetically modified seeds and prevented farmers from freely saving and replanting them without a proper licensing agreement.

Farmers' Seed Network v. Monsanto (2011) highlighted the ongoing debate surrounding intellectual property rights and the tension between protecting proprietary technology and preserving traditional farming practices. The case shed light on the challenges faced by farmers in an increasingly complex agricultural landscape, where patented seeds and restrictive licensing agreements have become more prevalent.

Questions raised by this case include:

1. Should farmers have the right to save and replant genetically modified seeds without infringing on patent rights?

2. How can a balance be struck between protecting intellectual property rights and preserving traditional agricultural practices?
3. What are the implications of restricting farmers' ability to save and replant patented seeds for seed diversity and resilience in agriculture?
4. Are there alternative models or approaches that can address the concerns of both seed developers and farmers regarding seed saving and exchange?

The outcome of this case reaffirmed the prevailing legal framework that upholds patent rights for genetically modified seeds, but it also sparked further discussions about the future of agricultural innovation, seed sovereignty, and the rights of farmers in relation to proprietary seed technologies.

References

Act of Parliament no. 53 of 2001, No. DL-33004/2000 (New Delhi, 30 October 2001), http://agricoop.nic.in/PPV&FR% 20Act,%202001.pdf.

Akanksha Tiwari, Piyusha Singh, Dhirendra Kumar Singh and Prabhat Tiwari(2022)Role of Intellectual Property Rights in Agriculture, Just Agriculture,Vol. 3 Issue-2.

Babu Rao (2019), "Issues and Challenges Of Intellectual Property Rights In Indian Agriculture",International Journal of Advanced Research, 7(2), 91-96. Article DOI: 10.21474/IJAR01

Debdyuti Datta, K.S. Das , D. Ghosh and S. K. Mondal(2016)," Intellectual Property Rights (IPR) in Agriculture: Prospects and Perspectives", National Academy of Agricultural Science (NAAS) Vol. 34, No. 4.

https://plantauthority.gov.in/

https://vikaspedia.in/agriculture/policies-and-schemes/crops-related/protection-of-plant-varieties-and-rights-of-farmers/protection-of-plant-varieties-and-farmers-rights-act2001#:~:text=Objectives%20of%20the%20PPV%20%26%20FR%20Act%2C%20 2001,To%20establish%20an&text=To%20recognize%20and%20protect%20 the,development%20of%20new%20plant%20varieties.

https://www.drishtiias.com/daily-news-analysis/the-protection-of-plant-varieties-and-farmers-rights-act

https://www.wipo.int/treaties/en/registration/pct/#:~:text=The%20Patent%20Cooperation% 20Treaty%20(PCT,of%20a%20PCT%20Contracting%20State.

Integrating Intellectual Property Rights and Development Policy, London.

Intellectual Property Rights: An Overview and Implications in Indian Markets (legalserviceindia.com)

Kalpana Sastry, R (2017). Intellectual Property Rights In Indian Agriculture, 105 FoCARS Foundation Course For Agricultural Research Service, Digital Repository of Course Materials, ICAR-National Academy of Agricultural Research Management.

Ministry of Law, Justice and Company Affairs (India), The Protection of Plant Varieties and Farmers' Rights Act, 2001,

Raj Chengappa, Patents: India wins a victory over turmeric but the war is on, India Today, Sept. 08, 1997, Updated on May. 14- Impact of Intellectual Property on the Agriculture Sector in India - Black n' White Journal (bnwjournal.com)

Ravishankar A & Archak S, Intellectual property rights and agriculture technology: Interplay and implications for India, Economic and Political Weekly, 35 (27) (2000) 2446-2452

Ravishankar and Archak (2000) Intellectual Property Rights and Agricultural Technology: Interplay and Implications for India. Economic and Political Weekly. Vol. 35, No. 27, pp. 2446-2452.

S.P, Subash, Gerard, Manju, Srinivas, Kondapi, Sastry and Kalpana. (2017). Developing Intellectual Property and Technology Management Professionals in NARES: The case of NAARM. 10.13140/RG.2.2.36742.88648.

Theresa M. Bevilacqua, Partner, and Kristin Stastny, Associate, Dorsey and Whitney LLP, Minneapolis, USA (2013), In the Courts: Monsanto v. Bowman: Supreme Court upholds patent holders' rights,WIPO Magazine. Agreement on Trade-Related Aspects of Intellectual Property Rights, Marrakech, 15 April 1994. The benefits of Intellectual Property Rights (tradecommissioner.gc.ca) What Role Does Intellectual Property Law Play in Agriculture? | LinkedIn

4

Corporate Governance and Sustainability Reporting in Indian Agriculture Context

Anita Bobade

I/C Director & Professor, DES's NMITD

"While directly serving their customers, companies should indirectly serve the interests of society by taking responsibility for the holistic impact of their activities. Its simply a more broad view of value creation. This is what ESG is all about." — *Hendrith Vanlon Smith Jr, CEO of Mayflower-Plymouth*

Learning Objectives

1. To understand the global best practice knowledge in corporate governance environment, the corporate forms, proper corporate, legal, and regulatory frameworks necessary to perform the duties of the board in protecting the rights of the shareholders.
2. The Evolution and History of Corporate Governance
3. To sensitize the students to the different models of Corporate Governance in India and abroad
4. To understand the importance of Global Reporting Initiatives and Explanation of the importance of corporate governance and sustainability reporting in the context of Indian agriculture and NAHEP goals

1.1. Introduction

Corporate governance while not a new concept, has, become an issue of global importance. Corporate governance is the set of processes, customs, policies, laws and institutions affecting the way a corporation is directed, administered or controlled. Corporate governance also includes the relationships among the many players involved (the stakeholders) and the goals for which the corporation is governed. The principal players are the shareholders, management and the board of directors. Other stakeholders include employees, suppliers, customers, banks and other lenders, regulators, the environment and the community at large.

"Corporate Governance is concerned with holding the balance between economic and social goals and between individual and communal goals. The corporate governance framework is there to encourage the efficient use of resources and equally to require accountability for the stewardship of those resources. The aim is to align as nearly as possible the interests of individuals, corporations and society." – Sir Adrian Cadbury. Corporate governance is typically perceived by academic literature as dealing with "problems that results from the separation of ownership and control". From this perspective, corporate governance would focus on: The internal structure and rules of the board of directors; the creation of independent audit committee's rules for disclosure of information to shareholders and creditors; and control of the management. he following illustration explains how a corporation is structured:

1.2. Need For Corporate Governance

Corporate governance therefore calls for three factors:

(1) Transparency in decision-making.

(2) Accountability which follows from transparency because responsibilities could be fixed easily for actions taken or not taken, and

(3) The accountability is for the safeguarding the interests of the stakeholders and the investors in the organization Implementation of corporate governance has depended upon laying down explicit codes, which enterprises and the organisations are supposed to observe.

1.2.a Objectives Of Corporate Governance

(1) To build an environment of trust and confidence amongst those having competing and conflicting interest.

(2) To enhance shareholders value and protect the interest of other stakeholders by enhancing the corporate performance and accountability.

(3) To have systems and procedures which are transparent and which can inform all the stakeholders about the working of the corporation.

1.2.b Definitions Of Corporate Governance

The concept of corporate governance sounds simple and unambiguous, but when one attempts to define it and scan available literature to look for precedence, one comes across a bewildering variety of perceptions behind available definitions. The definition varies according to the sensitivity of the analyst, the context of varying degrees of development and from the standpoint of academics versus corporate managements. However, there is an underlying uniformity in the thinking of all analysts that there is definite need to eradicate corporate misgovernance and promote corporate governance at all costs. It is not only the stakeholders who are keenly interested in ensuring adoption of best governance practices by corporates, but all societies and countries worldwide.

"Corporate governance is a field in economics that investigates how to secure/motivate efficient management of corporations by the use of incentive mechanisms, such as contracts, organizational designs and legislation. This is often limited to the question of improving financial performance, for example, how the corporate owners can secure/motivate that the corporate managers will deliver a competitive rate of return."

–www.encycogov.com, Mathiesen, 2002.

"The primary purpose of corporate leadership is to create wealth, legally and ethically. This translates to bringing a high level of satisfaction to five constituencies – customers, employees, investors, vendors and the society – at large. The raison d'être of every corporate body is to ensure predictability, sustainability and profitability of revenues year after year."

–N. R. Narayan Murthy

"Corporate governance deals with the ways in which suppliers of finance to corporations assure themselves of getting a return on their investment."

– The Journal of Finance, Shleifer and Vishny, 1997.

"Corporate governance - which can be defined narrowly as the relationship of a company to its shareholders or, more broadly, as its relationship to society…."

– From an article in Financial Times, 1997.

1.2.c Principles of Corporate Governance

Key elements of good corporate governance principles include honesty, trust and integrity, openness, performance orientation, responsibility and accountability, mutual respect, and commitment to the organisation. Of importance is how directors and management develop a model of governance that aligns the values of the corporate participants and then evaluate this model periodically for its effectiveness. In particular, senior executives should conduct themselves honestly and ethically, especially concerning actual or apparent conflicts of interest, and disclosure in financial reports.

Commonly accepted Principles of Corporate Governance

(1) Rights and Equitable Treatment of Shareholders: Organisations should respect the rights of shareholders and help shareholders to exercise those rights. They can help shareholders exercise their rights by effectively communicating information that is understandable and accessible and encouraging shareholders to participate in general meetings.

(2) Interests of Other Stakeholders: Organisations should recognise that they have legal and other obligations to all legitimate stakeholders.

(3) Role and Responsibilities of the Board: The board needs a range of skills and understanding to be able to deal with various business issues and have the ability to review and challenge management performance. It needs to be of sufficient size and have an appropriate level of commitment to fulfill its responsibilities and duties. There are issues about the appropriate mix of executive and non-executive directors. The key roles of chairperson and CEO should not be held by the same person.

(4) Integrity and Ethical Behaviour: Organisations should develop a code of conduct for their directors and executives that promotes ethical and responsible decision making. It is important to understand, though, that systemic reliance on integrity and ethics is bound to eventual failure.

(5) Disclosure and Transparency: Organisations should clarify and make publicly known the roles and responsibilities of board and management to provide shareholders with a level of accountability. They should also implement procedures to independently verify and safeguard the integrity of the company's financial reporting. Disclosure of material matters concerning the organisation should be timely and balanced to ensure that all investors have access to clear, factual information.

1.3. History and Evolution of Corporate Governance in UK

The Cadbury Report, formally entitled 'The Report of the Committee on the Financial Aspects of Corporate Governance' was published in December 1992, following the recommendations of the Cadbury Committee. The establishment of the Committee in May 1991 by the Financial Reporting Council, the London Stock Exchange, and the accountancy profession arose in response to the occurrence of financial scandals in the 1980's involving UK listed Companies, which led to a fall in investor confidence in the ability of the company's financial reporting.

As such the Committee addressed the financial aspects of corporate governance and subsequently produced a Code of Best Practice, the provisions of which, in their belief, all boards of UK listed companies should comply with. Specifically, they recommended that listed companies should incorporate a formal statement into their Report and Accounts outlining whether or not they complied with each of the Code's provisions. In respect of areas of non-compliance an explanation of the reason was sought. Further to this, the Report recommended that the compliance statements made by the companies be reviewed by auditors prior to release of the Annual Report. The stress in the Cadbury committee report is on the crucial role of the board and the need for it to observe the Code of Best Practices. Its important recommendations include the setting up of an audit committee with independent members.

The key focus of the provisions of the Code of Best Practice primarily related to the composition of the Board of Director's, the appointment and independence of non-executive directors, the service contracts and remuneration of executive directors, and company's financial reporting and controls.

Some of the main recommendations made were as follows:

(a) The majority of non-executive directors should be independent of management and management must be free from other relationships.

(b) Non-executive directors should be appointed for specified terms;

(c) Service contracts should not exceed three years;

(d) Executive remuneration should be subject to the recommendations of a remuneration

(e) Committee made up entirely of non executive directors

(f) An audit Committee, comprising of at least 3 Non Executive Directors should be appointed.

Following publication of the Code the London Stock Exchange introduced a requirement into the Listing Rules requesting all companies to include a statement of compliance, or non-compliance, with the provisions, in their annual Report and Accounts. Furthermore, institutional investors and Investment banks urged those listed companies for which they provided sponsorship and advice to adopt the provisions. As a result many companies changed their governance procedures and conduct accordingly.

1.4. Indian Model of Corporate Governance

The Indian corporate are governed by the Company's Act of 1956 that follows more or less the UK model. The pattern of private companies is mostly that of closely held or dominated by a founder, his family and associates. The figure drawn below indicates how the corporate governance system works in India.

In terms of the legislative mechanisms, Indian government and industry constituted three committees to study corporate governance practices in the country and suggest measures for improvement based on what has globally recognized as best practices. Significantly, most of the recommendations of the three committees- the SEBI –appointed Kumar Mangalam Birla Committee (2000), the government-appointed Naresh Chandra Committee (2003) and the SEBI's Narayana Murthy Committee are remarkably similar to those of England's Cadbury Committee and America's Sarbanes-Oxley Act, in terms of their approaches and recommendations.

The thrust of legislative reforms suggested by these committees and subsequent legislative actions adopted, centre around the strengthening of external governance mechanisms. "A key area here includes grater transparency and independent scrutiny of corporate accounts that are made available to investors. This is in line with the Anglo-American model where shareholders influence through the exit option which is contingent upon reliable and accurate information provided by companies. Institutional reforms, including a strengthening of oversight committees and the development of serious fraud office, are further evidence of the drive to seek for external monitoring of corporate affairs. In terms of reforms to internal mechanisms such as board of directors, it is notable that again the recommendations are centered on Anglo-American practice, namely, a greater role for Non-executive directors (NED's) and the curtailment of interlocking directorates".

Further experts point out that India has adopted the key-tenants of the Anglo-American external and internal control mechanisms, in the wake of economic liberalization and its integration into the global economy. "This is evident especially in the realm (kingdom, land) of the legislative framework where Indian policy-makers have taken their cue from UK and US committees and their recommendations." These various governance systems form a package of overall corporate control in each company law jurisdiction. It is vital to see the package as a whole. There has to be an integrated harmony between state legislation and regulatory infrastructure, stock market regulation and corporate self-regulation. Moreover, the overall corporate governance package has to be consistent with the way the business is done and the reality of relationship in that culture.

A Comparative Study of the Various Models of Corporate Governance

Sr. No.	Features	Anglo-American CG	German CG	Japanese CG	Indian CG
1.	Corporate objective.	Shareholder value.	Long-term corporate value.	Long-term corporate value.	Shareholder value.
2.	Government focus.	Capital Market.	Corporate body.	Keriestu or business network.	Maximize surplus.
3.	Shareholding.	Diffused institutional investors, significant block holders.	Banks, promoter families, other corporate.	Financial, non-financial corporate.	Directors and relatives. Other corporate, foreign investors.
4.	Measure of success.	Return on financial capital.	Return on human capital.	Return on social capital.	Return on financial capital.
5.	Decision making.	Checks and balances between voice and exit options.	Within the network of stakeholders includingemployees local community.	Within the network-includes business associates and banks as stakeholders.	Management, outside stakeholders excluded.
6.	Control ofcorporate.	Separated fromownership.	Linked with ownership.	Linked with ownership.	Linked with ownership.
7.	Orientation.	Short-term driven by market prices.	Long term.	Long term.	Short-term gains.
8.	Long-term investment in	Physical capital, R&D human capital.	Plant and equipment, employee training.	R&D, employee training.	Physical capital.
9.	Capital market (primary).	Liquid.	Less important, due toclose ties with banks.	Less important, due to close ties with banks.	Less important due o institutional funding.
10.	Capital market (secondary).	Important, frequenthostile takeovers possible.	Not important, hostile takeovers rare.	Not important, hostiletakeovers rare.	Not important, hostile takeovers rare.
11.	Investor commitment.	Low.	High, important in difficult times.	High, important in difficult times.	Low.
12.	Major investors.	Institutional shareholders, individualshareholders, business network, employees.	Banks, Government, individual shareholders and Institutional shareholders.	Government, individual shareholders and Institutional shareholders, main bank.	Directors and relatives other corporate, foreign investors, Institutional investors (UTI)

Sr. No.	Features	Anglo-American CG	German CG	Japanese CG	Indian CG
13.	Board composition.	Executive and non-executive directors.	Two-tier boards (supervisory board andmanagement board).	Executive and non-executive directors (representing outside finance institutions).	Executive and non-executive directors.
14.	Goal of the board.	To promote shareholder wealth.	To promote long temorganizational health.	To promote long temorganizational health.	Shot0term gains.
15.	Board independence over management.	Little.	High.	Little formally, more formally.	Little.
16.	Executive composition.	High.	Moderate.	Low.	Moderate, subject to government approval.
17.	Dividend.	High.	Low.	Low.	Low, uncertain.
18.	Strength.	Dynamic, market-based, liquid capital, internalization non-problematic.	Long-term industrial strategy, stable capital,strong overseas investment.	Long-term industrial strategy, stable capital.	Recent government and organizational activism (CII) towards corporate governance practices.
19.	Weakness.	Instability, short-termism.	Internationalization difficult; vulnerable toglobal capital market.	Secretive, corrupt practices, growth in institutional activism and financial speculation in recent times.	Lack of proper disclosure; secretivecorrupt practices; instabilities.

1.4.a. Corporate Governance – Anglo Saxon (American) Model

(1) Ownership is more or less equally divided between individual shareholders and institutional shareholders.

(2) Companies are run by professional managers, who have negligible stakes.

(3) Directors are not independent of managers.

(4) Short term orientation.

(5) Planning a proactive role in corporate governance.

(6) Disclosure norms are comprehensive, rules against insider trading tight and penalties for price manipulation stiff.

(7) Takeover code not very comprehensive.

(8) High tension model – Important role of the CEO, active capital market, short term orientation and takeover the eats.

1.4.b. Corporate Governance – German and Japanese Model

(1) Banks and financial institutions have substantial stakes in the equity.

(2) View themselves as long term investors.

(3) Disclosure norms are not very stringent, checks on insider trading are not very comprehensive and effective.

(4) No market for corporate control in Germany and Japan. Takeovers or hostile takeovers are very rare.

(5) Networked model – Access to vast information through banks, long term orientation, infrequent mergers and acquisitions and limited role of capital market.

(6) KEIRETSU – in Japan.

1.4.c. Corporate Governance – India – Private Sector

(1) Three broad categories of shareholders, promoters and financial institutions including MFs and individual investors.

(2) Majority rule voting system.

(3) Board consists of promoter director, professional director, institution nominee directors.

(4) Candour and forthrightness sacrificed at the altar of courtesy and politeness.

(5) Nominee directors do not have any incentive to monitor behaviour and performance of management seriously.

(6) Scattered and ill organised individual shareholders are not able to accomplish much.

(7) Family Managed Businesses display greater entrepreneurial vigor, act more proactively and exercise stricter control.

(8) React slowly to view opportunities and put greater emphasis on systems.

(9) Entrenched system given the firm hold of promoters and the disinclination and inability of others to challenge them.

1.4.d. Corporate Governance – India – Public Sector

(1) Equity shares are owned wholly or substantially by government.

(2) Three categories of directors – financial, government directors, outside directors.

(3) Political and bureaucratic influence owes management of a PSU.

(4) Constrained by various regulations and administrative guidelines. CAG audit and accountability to parliament.

(5) Chief executives have short tenure.

(6) Performance standards are soft, compensation levels low, incentives for performance poor and real accountability is weak.

1.4.e. Corporate Governance – India – Co- Operative Sector

(1) Equity shares are owned wholly or substantially by government.

(2) Political and bureaucratic influence on the management of the co-operative societies and banks

(3) Constrained by various regulations and administrative guidelines.

(4) Chief executives Officer or Managing Director does not have complete autonomy

(6) Performance standards are soft, compensation levels low, incentives for performance poor and real accountability is weak.

1.5. Parties to Corporate Governance in A Co- Operative Society/Bank

Parties involved in corporate governance include the governing or regulatory body (e.g. the Securities and Exchange Commission in the United States), the Chief Executive Officer, the Board of Directors, Management and Shareholders. Other stakeholders who take part include suppliers, employees, creditors, customers and the community at large. In cooperatives society, the principal (shareholder) delegates decision rights to the agent (manager) to act in the principal's best interests. This separation of ownership from control implies a loss of effective control by shareholders over managerial decisions. Partly as a result of this separation between the main two parties, a system of corporate governance controls is implemented to assist in aligning the incentives of managers with those of shareholders, in order to limit the self-satisfying opportunities for managers. With the significant increase in equity holdings of institutional investors, there has been an opportunity for a reversal of the separation of ownership and control problems because ownership is not so diffuse.

A board of directors often plays a key role in corporate governance. It is their responsibility to endorse the organization's strategy, develop directional policy, appoint, supervise and remunerate senior executives and to ensure accountability of the organization to its owners and authorities.

All parties to corporate governance have an interest, whether direct or indirect, in the effective performance of the organization. Directors, workers and management receive salaries, benefits and reputation; whilst shareholders receive capital return. Customers receive goods and services; suppliers receive compensation for their goods or services. In return these individuals provide value in the form of natural, human, social and other forms of capital.

A key factor in an individual's decision to participate in an organization (e.g. through providing financial capital or expertise or labor) is trust that they will receive a fair share of the organizational returns. If some parties are receiving more than their fair return (e.g. exorbitant executive remuneration), then participants may choose to not continue participating...potentially leading to organizational collapse (e.g. shareholders withdrawing their capital). Corporate governance is the key mechanism through which this trust is maintained across all stakeholders.

1.6. Corporate Governance in Indian Co-operative Banks

The corporate governance practices are widely used by the banking institutions. Corporate Governance would study issues like protection of shareholders' rights, enhancing shareholders value, board issues including composition and role, disclosure requirements, integrity of accounting practices, the control systems, in particular internal control systems.

Corporate governance especially in the cooperative sector has come into sharp focus because more and more cooperative banks in India, both in urban and rural areas, have experienced grave problems in recent past which has in a way threatened the profile and identity of the entire cooperative system. The cooperative banks are having a history of more than a century and the governance problems including mismanagement, financial impropriety, poor investment decisions and growing distance between the members and their cooperative societies have surfaced.

Cooperatives are organised by groups of people and jointly managed and democratically controlled enterprises. They exist to serve their members and depositors and produce benefits for them. Cooperative corporate governance is therefore about ensuring cooperative relevance and performance by connecting members, management and employees to the policy, strategy and decision making processes. The corporate governance practices are generally helps in improving the functioning of the cooperative banks, including the internal & external control aspects, board transparency, accountability, shareholders protection and disclosure aspects of the cooperative banks. Co-operative societies are organised groups primarily working at grass root level serving for the financially weaker section of the society, who have limited access to the Banks. The success of the co-operative sector lies in the governance of the society and it is in the hands of the elected representatives of the shareholders. Shareholders are the real owners of the society, but in real terms society is managed by 10-15 elected members of the Board. Shareholders normally come once in a year at annual general meetings and are unaware about the day-to-day affairs of the society they get information only through annual report which is made available to them at the time of annual general meeting. Hence there is a need to include maximum information about Board of directors and its functions (administrative as well as financial) in annual report to bring transparency in the business to protect the interest of various stakeholders.

By defining board procedure, code of conduct for board members and by disclosing all the information relation to financial and administrative matters in annual report and website co-operative sector can achieve good governance. Good governance for co-operative sector is much needed to protect the interest of various stakeholders' in co-operative sector. If every society and its board of members take responsibility of improving governance of society then, definitely co-operative sector will reach greater heights. The Cooperative values are focused on transparency, responsibility, solidarity and sustainability. Board members have a stewardship role, a responsibility towards members and stakeholders to shaping the bank and improving the financial health through corporate governance. The long term and short term interest of stakeholders have to be managed effectively.

1.7. The Global Reporting Initiative (GRI) is an organization that produces a comprehensive sustainability reporting framework that is widely used around the world. It started with a vision of a sustainable global economy where organizations manage their economic, environmental, social and governance performance and impacts responsibly and report transparently. GRI has a mission to make sustainability reporting standard practice by providing guidance and support to organizations. It's about disclosure of organization's economic, environmental, social and governance performance in credible way and subsequently helping them to identify business risks and opportunities to the organization due to these parameters.

GRI provide framework to report organization's sustainability performance. It is known as the sustainability reporting framework and consists of three parts

(1) Sustainability Reporting Guidelines,

(2) Sector Supplements and

(2) The Technical Protocol – Applying the Report Content Principles. All it provide guidance on how organizations can disclose their sustainability performance in true manner. Organization of any size and type can use this framework to assess their performance indicators in economic, environmental, and social domain. Organization that creates their non financial report based on GRI sustainability reporting framework can be considered as credible output aligned with applicable standards and norms. Such report can be used to demonstrate organizational commitment to sustainable development. It can also qualitatively and quantitatively used to measure and compare organization performance based on environment, social and economic dimensions.

1.7.a. Background on Sustainability Reporting in 1987, the World Commission on Environment and Development set out an aspirational goal of sustainable development - describing it as "development which meets the needs of the present without compromising the ability of future generations to meet their own needs.' Through their activities and relationships, all organizations make positive and negative contributions toward the goal of sustainable development. Organizations therefore have a key role to play in achieving this goal. Sustainability reporting is an organization's practice of reporting publicly on its economic, environmental, and social impacts, and hence its contributions - positive or negative -towards the goal of sustainable development. Sustainability reporting has become a mainstream business practice. Almost three-quarters of N100 companies now report on corporate responsibility, according to KPMG's 2015 Survey of Corporate Responsibility Reporting. Of the world's largest companies - the G250 - 92% produce sustainability reports, with 74% of these referring to the GRI Guidelines or Standards. Globally-accepted standards, such as the GRI Sustainability Reporting Standards, provide a common language and credible set of disclosures for organizations to communicate about their impacts on the economy, the environment, and society. Organizations can combine their use of such instruments to improve the quality and comparability of information they report publicly. This information enables better decision making by organizations and their stakeholders, helping to build the long-term trust that is essential for the functioning of markets

1.7.b. NAHEP

The Govt. of India (GoI) has received financing in the amount of US$ 82.5 million from the International Bank for Reconstruction and Development (IBRD) for funding the National Agricultural Higher Education Project (NAHEP) of the Indian Council of Agricultural Research (ICAR). The total cost of the project is estimated at US$ 165 million. It is proposed to apply part of the proceeds for eligible payments towards Consultancy Services for Consultant Training for Faculty, Student and Staff of PIU to be hired for providing support to the Project Implementation Unit (PIU) of the NAHEP,

1.7.c. ICAR

NAHEP will support the education development of Agricultural Universities (AUs) including State Agricultural Universities, Central Agricultural Universities and Central Universities with Agricultural Faculty, Deemed-to-be Universities and ICAR. Objective of the NAHEP is to further improve policy, quality assurance through accreditation, common academic regulations and governance, effective curricula delivery systems, improvement of faculty competence, student development, attracting talented students, IT support and up-gradation of infrastructure and facilities in enabling the system to catch up nationally and internationally.

1.7.d. GRI, or Global Reporting Initiative, sets standards for companies to report on their economic, environmental and social impact. SASB, or

Sustainability Accounting Standards Board, focuses specifically on the financial materiality of sustainability issues.

The GRI Standards represent global best practice for reporting publicly on a range of economic, environmental and social impacts. Sustainability reporting based on the Standards provides information about an organization's positive or negative contributions to sustainable development.

The GRI Standards is a modular system of interconnected standards. Three series of Standards support the reporting process: the GRI Universal Standards, which apply to all organizations; the GRI Sector Standards, applicable to specific sectors; and the GRI Topic Standards, each listing disclosures relevant to a particular topic. Using these Standards to determine what topics are material (relevant) helps organizations to achieve sustainable development.

GR13, Agriculture, Aquaculture and Fishing Sector, 22 would come into effect from 1st Janaury, 2024.

Chapter Section Summaries

Section 1. The first section aims to introduce students to the concept of corporate governance, with a focus on best practices globally. Readers will learn about the various corporate forms, legal obligations, and regulatory frameworks that are necessary for a board to effectively protect the rights of shareholders.

Section 2. The second section explores the evolution and history of corporate governance, highlighting how the field has developed over time, the factors driving the evolution of corporate governance, and key moments in corporate governance history.

Section 3. The third section sensitizes students to different corporate governance models in India and abroad, examining how different systems work to protect the rights of shareholders, and what key differences exist between models.

Section 4. The final section focuses on the importance of sustainability reporting and how it relates to corporate governance and NAHEP goals within the Indian agriculture sector. Students will learn about the Global Reporting Initiative and how companies can use this framework to improve their sustainability reporting practices. The section will also emphasize the importance of aligning sustainability reporting with corporate governance frameworks and the benefits this can bring to both organizations and stakeholders alike.

Discussion Questions

1. What are the key components of a proper corporate governance environment?
2. How has corporate governance evolved over time and what has driven this evolution?
3. How can the principles of corporate governance be applied in the Indian context and why is it important to do so?
4. What are the key differences between various models of corporate governance, and how do they impact the rights of shareholders?
5. What are some potential weaknesses or challenges within the current corporate governance structure in India, and how can they be addressed?
6. How can a company create an effective sustainability reporting system that aligns with their corporate governance framework, and what benefits can this bring to both the organization and the broader stakeholder community?

Application Case Study

Ethical Managers Make Their Own Rule

In 1900 Queen Victoria sent a decorative tin with a bar of chocolate inside to all her soldiers who were serving in South Africa. These tins still turn up now after almost 2 centuries which is indeed a tribute to the collective instincts of the British.

In 1824, John Cadbury began vending tea, coffee, and drinking chocolate, which he produced himself, at Bull Street in Birmingham, England. John Cadbury later moved into the production of a variety of Cocoas and Drinking Chocolates being manufactured from a factory in Bridge Street, supplying mainly to the wealthy due to the high cost of manufacture at this time. During this time a partnership was struck between John Cadbury and his brother Benjamin. At this time the company was known as 'Cadbury Brothers of Birmingham'.

The two brothers opened an office in London and in 1854 received the Royal Warrant as manufacturers of chocolate and cocoa to Queen Victoria. Around this time in the 1850s the industry received a much needed boost with the reduction in high import taxes on cocoa; this allowed chocolate to become more affordable to everyone.

Due to the popularity of a new expanded product line, including the very popular Cadbury's Cocoa Essence, the company's success led to the decision in 1873 to cease the trading of tea. Around this time, master confectioner Frederic Kinchelman was appointed to share his recipe and production secrets with Cadbury, which led to an assortment of various chocolate covered items.

In 1878, John Cadbury's sons Richard and George (who had taken over the company after John Cadbury's retirement in 1861), acquired the Bournbrook estate, comprising fourteen and a half acres of countryside five miles south of the outskirts of Birmingham. They renamed the Bournbrook estate to Bournville and opened the Bournville factory in 1879. In 1893, George Cadbury bought 120 acres of land close to the works and planned, at his own expense, a model village which would 'alleviate the evils of modern more cramped living conditions'. By 1900 the estate included 313 cottages and houses set on 330 acres (130 ha) of land. As the Cadbury family were Quakers there were no pubs in the estate; in fact, it was their Quaker beliefs that first led them to sell tea, coffee and cocoa as alternatives to alcohol.

When Queen Victoria summoned Mr. George Cadbury to give the chocolate order he was faced with an ethical dilemma – Cadbury then was the second largest chocolate manufacturing company in Britain and was aspiring to be the number ONE. Yet he himself opposed the Anglo Boer War. He was able to resolve the Dilemma by Accepting The Order And Fulfilling It But Without Earning Any Profit. He felt the war was unjust and he did not support the war so he did not earn the profits from the chocolates given to the British soldiers. He also was able to generate employment and wealth for his workers and hence all were happy.

Mr. George was able to do this as he was a sole proprietor then and he could compromise his personal ethics for the good of the business and his employees. However when there are millions of shareholders and a very well diversified company how one does manage these dilemmas?

Questions

1. Define Corporate Governance.
2. What are the objectives and principles of corporate governance
3. Trace the history and evolution of corporate governance in UK.

4. Make a comparative study of the different models of corporate governance and highlight the significant features of Corporate Governance for Co-operative Society and Bank
5. What is the need of Corporate Governance in Co-operatives in India

References

Afsharipour, a Brief Overview of Corporate Governance ReformsIn India, UC Davis Legal Studies Research Paper Series. April 2011. www.ijcrt.org © 2018 IJCRT | Volume 6, Issue 2 April 2018 | ISSN: 2320-2882 IJCRT1893330 International Journal of Creative Research Thoughts (IJCRT) www.ijcrt.org 282

Chhibber, P.K., and S.K. Majumdar (1999), Foreign Ownership and Profitability:Property Rights, Control, and the Performance of Firms in Indian Industry, Journal of Law and Economics.

Das, A. and S. Ghosh, 2004. Corporate Governance in Banking System: An Empirical Investigation, Economic and Political Weekly, Volume 2, 2004.

Dimple Grover, Amulya Khurana, Ravi Shankar, the Regulatory NormsOf Corporate Governance in India.

Goswami, Omkar, 2002, "Corporate Governance in India," Taking Action Against Corruption in Asia and the Pacific. https://icmai.in/upload/Institute/Updates/SUKHINOBHAVANTU_SSB_NEWSLETTER_FEBRUARY_2023_Vol1.pdf

https://www.globalreporting.org/how-to-use-the-gri-standards/gri-standards- english-language/ https://www.globalreporting.org Ready to Report ? GRI Linking the GRI Standards and the SEBI BRR Framework, Linkage document SEBI Notifications on BRR https:// www. caclubindia.com/articles/global-reporting-initiative-gri-empowering-sustainable-decisions-29752.asp

M. L. Sukhdeve (2012) Corporate Governance in Cooperative Banks-Best Practices in Loan & Pricing Mechanism - The Maharashtra State Coop. Bank Ltd. Mumbai https://bird-cpec.nabard.org/wp-content/uploads/2018/01/Best-Practices-1-Sukhdeve-Sir-Loan-Pricing-paper.pdf

Mallikarjun Mardi, "Corporate Governance Practices in Selected Co-Operative Banks at Vijayapur City" https://www.researchgate.net/publication/324173730

SadhalaxmiVivek Rao, Legal Framework and corporate Governance:An Analysis of Indian Governance System. Voluntary Guidelines on Corporate Governance, the Ministry of Corporate Affairs, December 2009.

Annexures

Annexure 1: Recap of Key Points Covered in the Volume

This volume has explored the need for sustainable growth and development in the Indian agriculture sector, given the environmental, social, and economic challenges it faces. The chapter has highlighted that implementing sound environmental, social, and governance policies, management systems & processes, corporate governance, and intellectual property rights can contribute significantly to achieving sustainability in the sector. Additionally, the chapter has discussed the specific role of the National Agricultural Higher Education Project (NAHEP) in promoting sustainable growth through enhancing higher education in the sector.

Annexure 2: How Implementing Environmental, Social, and Governance Policies, Management Systems, and Intellectual Property Rights Can Lead to Long-Term Sustainable Growth and Development in Indian Agriculture?

Implementing environmental, social, and governance policies, management systems & processes, corporate governance, and intellectual property rights can help attain long-term sustainable growth and development in Indian agriculture. These measures play a vital role in ensuring that agriculture practices are sustainable and contribute to preserving the environment, enhance social welfare and governance while ensuring transparent and ethical farming practices. Proper environmental policies ensure that natural resources are conserved, leading to their sustainable use for generations. Policies for social and governance ensure that farmer well-being is a top priority, with effective decision-making practices, while intellectual property rights provide incentive to innovate new sustainable farming practices.

Annexure 3: Call-to-Action for Organizations to Take Action on These Important Topics in the Context of NAHEP and Indian Agriculture

Indian agriculture has experienced significant growth in recent decades; however, sustaining that growth is challenging, given the sector's unique challenges. In the context of the National Agricultural Higher Education Project (NAHEP), organizations involved in the sector must take action to promote sustainable development. This includes the implementation of sound environmental, social, and governance policies, management systems & processes, corporate governance policies, and promoting intellectual property rights. The establishment of these initiatives within the context of NAHEP will lead to effective policymaking, bring efficiency in agricultural practices and contribute significantly to attaining sustainable growth. Furthermore, organizations must collaborate and commit to enhancing the quality of higher education in the agricultural sector, as this will shape and encourage future sustainable farming practices.

In conclusion, the Indian agricultural sector is facing several sustainability challenges that threaten its long-term growth and development. Organizations and individuals involved in the sector must take action to promote sustainable development, and initiatives like NAHEP play a critical role. The implementation of sound environmental, social, and governance policies, management systems & processes, corporate governance policies, and promoting intellectual property rights can achieve sustainability in the agriculture sector. Furthermore, a commitment to enhancing the quality of agricultural education through the NAHEP initiative is critical in promoting sustainable growth in the Indian agriculture sector. Organizations must act now to achieve sustainable development goals in the Indian agriculture sector.

FSC
www.fsc.org
MIX
Papier | Fördert gute Waldnutzung
FSC® C083411

Zeitfracht Medien GmbH
Ferdinand-Jühlke-Straße 7
99095 Erfurt, Deutschland
produktsicherheit@kolibri360.de